89

THE HISTORY OF THE STEEL HELMET IN THE FIRST WORLD WAR

Also by Haselgrove and Radovic
Helmets of the First World War: Germany, Britain & their Allies

Also by Radovic
German Helmets of the Second World War (*two volumes*)
German Camouflaged Helmets of the Second World War (*two volumes*)

THE HISTORY OF THE STEEL HELMET IN THE FIRST WORLD WAR

VOLUME ONE

• AUSTRO-HUNGARY • BELGIUM • BULGARIA •
• CZECHOSLOVAKIA • FRANCE • GERMANY •

MICHAEL J. HASELGROVE
& BRANISLAV RADOVIC

Schiffer Military History
Atglen, PA

Acknowledgments

The authors wish to thank the following people who have contributed to the preparation and content of this book by providing photographs, information or access to their private collections. We have, wherever possible, given credit to the owners of the photographs or helmets that appear in the book. We are particularly grateful to the following for their generosity in sharing their knowledge and/or precious possessions with us: David Acon, Michael Baldwin, Jean Marc Barris, Alan Beadle, John Burnett, Stephen Chambers, Marcus Cotton, Malcolm Fisher, Daniel Fisher, Brian Maederer, David Powers, Greg Nevill, Tony North, Andy Read, Ellwood von Seibold, Robert Tredwen, Matthew Tredwen, Charles Woolley.

In addition we would like to thank the following institutions and their staff:

The Imperial War Museum, London and Duxford:
Martin Boswell, Diana Condell, Alan Cant, Tom Duplock, Helena Nicholls

The Polish Army Museum, Warsaw:
Director Jacek Macyszyn, Wojciech Potrzebnicki, Zdzislaw Zajdler

Le Musée Militaire du Perigord, France:
President Jacques Dornier, Jacques Moniotte, Claude-Jean Blanchard

All the above museums are a "must visit" for anyone interested in military history. The authors strongly recommend them to their readers.

The authors would be grateful to any of our readers who might have additional information, material or photographs that could enhance our collective knowledge of the subject and perhaps be included in any update of these volumes.

Lastly, we would like to thank our long-suffering wives for their assistance and great patience: Judy Haselgrove and Anna Chajkin.

Book design and photography by Robert Biondi and Branislav Radovic.

Library of Congress Catalog Number: 2006925465.

Printed in China.
ISBN: 0-7643-2528-0

We are always looking for people to write books on new and related subjects. If you have an idea for a book, please contact us at the address below.

Published by Schiffer Publishing Ltd.
4880 Lower Valley Road
Atglen, PA 19310
Phone: (610) 593-1777
FAX: (610) 593-2002
E-mail: Info@schifferbooks.com.
Visit our web site at: www.schifferbooks.com
Please write for a free catalog.
This book may be purchased from the publisher.
Please include $3.95 postage.
Try your bookstore first.

In Europe, Schiffer books are distributed by:
Bushwood Books
6 Marksbury Ave.
Kew Gardens, Surrey TW9 4JF
England
Phone: 44 (0)20 8392-8585
FAX: 44 (0)20 8392-9876
E-mail: info@bushwoodbooks.co.uk
www.bushwoodbooks.co.uk
Free postage in the UK. Europe: air mail at cost.
Try your bookstore first.

FOREWORD

Defensive armour, from its origins, was made from a variety of materials including metal, textiles, leather and whalebone. It was developed for use in warfare as a protection against swords, spears and missile weapons, principally arrows. Even in its earliest forms, the concept of parade was a high priority for the armourer. The highly embossed bronze helmets and cuirasses of Greece and Rome not only offered protection to the wearer but also reflected his wealth and status.

By the Eleventh Century in Europe, the most widely used defensive armour was mail. This consisted of simple garments constructed from a series of wrought-iron rings interlocked and usually riveted. It was worn over a padded garment of similar proportions to prevent the skin being chafed. This is the armour depicted in the Bayeux Tapestry of the 1080s. The lower section of the garment was split so it could be worn on horseback. Armourers were concerned, from early times, with protection of the head. The helmets shown in the Bayeux Tapestry are of conical shape with a distinctive nasal guard to protect the face. Existing helmets of this form show that they were constructed from a series of bands and plates. However, at least one helmet has survived which has been raised from a single sheet of iron. This is the helmet of St. Wenceslas in Prague Cathedral, which must date from before his death in 935AD.

Very few examples of early armour survive especially from the early medieval period. Thus, the development of armour has to be sought from a variety of sources. Monumental brasses, effigies, sculpture and contemporary manuscripts are the most important of these. The development of plate armour starts in about 1250AD. Initially, it was limited to defence of the knees and elbows but, by the early 15th century, a garment mounted with large metal plates to protect the body became increasingly used. There are references to metal gauntlets by the end of the 13th century and by the 1330s, plate armour was replacing mail. The gradual decline in the use of mail may have been due to the increased effectiveness of the longbow and especially in the development of long steel arrowheads, known as bodkins, which could successfully penetrate metal rings.

The 14th century saw the development of the breastplate. Surviving early examples show that armourers had quickly learnt the advantages of rounded surfaces and ribs to deflect arrows and lances. This is also reflected in the change in the shape of helmets, which by the 1300s had changes from flat-topped forms to the conical crown and pointed visor. The 14th century also saw the development of the so-called kettle-hat – a form that was to last up to the Second World War.

From about 1375, the history of armour can be seen from actual examples. Preserved in famous armouries, such as that of Schloss Churburg in the Tyrol, are some remarkable 14th century armours in fine condition. The great period of armourers is generally thought to be between 1410 and 1500, but it should also be noted that some very fine armour was made as late as the second half of the 17th century – especially in England.

Helmets with visors became increasingly common in the first half of the 14th century. The Great Helm worn, over the *bascinet*, is also frequently shown in manuscripts and on military effigies. Also in this period, armour specially designed for the joust began to appear. The jousting helm was characterised by its great depth and plain front.

The 15th century saw the development of the fully armoured knight. Every part of the body was covered with carefully shaped plates; the finest with applied and embossed decoration in the late-Gothic style. It was during this period that the great armour-making centres, such as Milan, Augsburg, Landshut and Nuremburg, became pre-eminent, each with their own particular style and character.

Armour was made from a series of flat plates, suitably shaped by hammering over a series of shaped stakes. Contemporary illustrations show the plates being worked cold but they must have been frequently annealed to allow them to be worked. Helmets were raised from flat plates, rather like a silversmith making a bowl. The best armours were carefully fitted to an individual, defensive armour for the legs being the most difficult to make.

In the last quarter of the 15th century, the garniture was invented, principally for tournaments. This consisted of a large series of separate pieces of armour all with matching decoration, which could be put together for different forms of combat, such as fighting on foot or jousting.

The 16th century witnessed the widespread use of armour designed especially for parade. Every form of decoration was used on armour. Gilding, engraving, etching, embossing and damascening were all used to make armour look as impressive as possible. These techniques were usually combined with blueing – a treatment that darkened the steel surface and, thus, heightened any gold or silver decoration.

The type of helmet known as a *sallet*, with its characteristic long projection at the back, had been developed in the latter part of the 15th century. This developed, in the 16th century, into the close-helmet, in which the cheek-pieces pivot at the side and are covered by a large visor.

The influence of the Renaissance was quickly reflected in armour. Celebrated Italian armourers, such as the Negroli and Piccinino families who worked in Milan, produced highly embossed armour in the Mannerist style – the form copying Greek and Roman armour.

Firearms began to dominate the battlefield in the first quarter of the 16th century. The battle of Bicocca in 1522 was won principally by the use of firearms. It was soon realised that the armour could be made bulletproof. As the armour became heavier, the various plates then began to lose their elegant shape.

Armourers had been concerned, in the 15th century, with making breastplates resistant to spears and arrows by using a specially developed quenching process to harden the steel. This process continued into the 17th century, with some breastplates being made of three thicknesses.

By the end of the 16th century, large parts of the harness were abandoned. Half-armours became common, as did very heavy siege helmets, which could only have been worn for short periods. Armour was still made for parade but the emphasis was on the applied decoration not the shape. Compared to the 15th century, harnesses looked ugly and ungainly.

With the arrival of the relatively reliable firearm, in about the middle of the 17th century, full plate armour was relegated from the practical, to the ceremonial. A typical *harquebeusiers* (musketeer) armour of the period consisted simply of a light helmet, a back and breastplate, and a single defence of the arm that held the reins of a horse. This lighter arrangement of cuirass continued into the first years of the Great War on the backs of French cavalrymen where it soon became clear that this form of body armour was now a thing of the past, it being neither proof against the bullet nor blast, convenient to wear, nor inconspicuous.

However when casualties from rock and metal fragments began to rise alarmingly in the first year of the war, the combatant countries turned, once again to their history books to reconsider the question of personal protection for the soldier, particularly for the head. Many of the proposals were entirely inappropriate, being too heavy and too difficult to produce in large quantities. Some were little more than copies of helmets from earlier epochs.

Plate armour in its crudest form was resurrected, but in the shape of trench loophole plates and body armour, worn under and over the uniform. The mace was re-introduced as the trench club, the cavalry horse replaced by the tank. A fundamental change in warfare was forced upon the world – the technology of death was advancing rapidly and by 1915 all the major warring nations were frantically striving to provide the soldier with a helmet that could reconcile the seemingly irreconcilable, namely weight over manoeuvrability and quality over critical production costs.

This book, in two volumes, illustrates and highlights those problems, charting as it does the early tentative steps towards a general recognition and full re-introduction of the protective helmet for warfare. Recently, in the last twenty years, the moulded kevlar helmet has replaced steel as the material of choice for most armies. From the earliest examples of 14th century BC boar tusk helmets from our Mediterranean pre-history to 21st century AD troops patrolling the streets of Baghdad and Basra, the principle remains the same.

Anthony R.E. North, AMA, FSA, FRSA, October 2005
Forty years as curator at the Victoria and Albert Museum, London, metalwork department, specialising in arms and armour.

CONTENTS · VOLUME ONE

CONTENTS · VOLUME TWO

INTRODUCTION

For those familiar with our first book – *Helmets of the First World War* – some explanation is necessary as to why we thought it appropriate to produce a further book on the same subject. That first book featured helmets solely from our respective collections at the time, together with contemporary photographs. We resisted the temptation to include photographs of helmets from other collections as we were aware that this would lead to an escalation of the project. However, this approach led to obvious gaps in the helmets covered, with some countries omitted completely.

This book, the reader will find, differs in a number of respects from our first book. Firstly, we have included photographs of helmets not just from our own collections but also from a number of other collections, including those of the Imperial War Museum, London and the Polish Army Museum, Warsaw. The collection or collector to which each helmet belongs is credited, with a few exceptions, such as where the present whereabouts of a given helmet is unknown. Secondly, each country represented has a separate section. Thirdly, at the beginning of each section we have included, wherever we thought it appropriate, some detail as to the involvement of the country in question in the war, together with details of the helmet(s) used by that country.

Commenting briefly on the above points, the reader will find that a few of the photographs of helmets used in our earlier book are repeated here. We have only done so where completeness required it, or where the photograph in question served to show a helmet particularly well. For example, those familiar with the earlier book will recall that we showed numerous photographs of German camouflage painted helmets, a feature repeated in this book. However, of all the helmets photographed for this book, only three or four appeared in the earlier book. They are included in this book because they are particularly fine or interesting examples.

Once we took the decision to have a separate section for each country, it was a matter of deciding which countries should be included. Major participating countries had to be included. For the rest, the existence and availability of an identifiable helmet used by that individual country determined inclusion. It is a fact that when, on the 28th June 1919, the Treaty of Versailles was signed, it was signed by a representative of one of each of the twenty-seven victorious countries, all of whom had declared war on Germany. The part played by some of those countries was, comparatively speaking, modest. In respect of participating countries not included in this book, it has been impossible to identify a helmet used by them during the war. Additionally, we have not included separately Canada, Australia, India, New Zealand, South Africa and others forming part of the former British Empire or Dominions, as those countries used the same helmets as British troops. Finally, one country that remained neutral, namely Holland, has been included. This is largely because of the interesting incursions into Dutch territory by British and German troops during the war.

In some cases we have given brief details of the involvement of the country concerned, in the war. Readers should not regard the details as comprehensive and those requiring more information must look elsewhere. Where details are given, they are simply to help to explain the involvement of that country in circumstances that may perhaps not be well known, or which are otherwise somewhat obscure. The details given may help to explain the supply and use of a particular helmet(s) where that source of supply might not otherwise be obvious.

We have also included in the introduction to each country, as much detail as is reasonably possible relating to the development, manufacture and introduction of the helmet(s) used. The information given comes from many sources and attention is drawn to the bibliography. In some cases, particularly where a given helmet was available for examination and measurement, we have been able to check the information given. In other cases original source material was not available or accessible and we have had to rely upon what has been written by others without the means to check the correctness or otherwise. We have not excluded such material upon the basis that some of it may be incorrect. Rather, it is included in the hope that debate is stimulated, thus expanding the knowledge of our subject. However, inasmuch as there are errors included in what we have written, we have done our best to exclude them. We can only apologise for those that have slipped through.

The reader will find in the introduction at the start of some sections comments about the use of a particular helmet after the end of the Great War. We are aware that some readers may consider such detail irrelevant. We comment that the information is included to provide an understanding of the use and modification to which some helmets were subject after the end of the war. The information may serve to explain why some helmets, manufactured in what seem to us today to be such staggering quantities, are now so rare in their original configuration.

Whereas in the text we have included some information about the helmets in use at the start of the war, to limit the size of the book, we have only included photographs of steel combat helmets and those designed to resemble them, such as lightweight parade helmets.

A major benefit of including helmets from other collections, apart from making the book more comprehensive, has been that we have been privileged to handle and photograph helmets, some of them extremely rare, to which we might not otherwise have had access. Our personal knowledge has thus been expanded and we hope that some of that knowledge is adequately passed to the reader.

Our initial plan to produce a single volume book proved unworkable, given the quantity of material to be covered. As work on the book progressed, we became increasingly aware that not only did we have a great deal of material to include, but that there must be a large amount of information still to be discovered. Whether information presently not available will be released as archives are opened to public scrutiny or is otherwise revealed, or whether information that must have existed has been lost or destroyed in the years since the end of the Great War, remains to be seen. What is certain is that we continue, as at the time of writing in September 2005, to receive newly discovered information and photographs from those individuals kind enough to provide them.

We have considered whether it is possible to draw any conclusions or correlation between the number of battlefield dead during the Great War and the introduction and use of steel protective helmets on a large scale. It is true to say that alarm at the number of head wounds prompted development and introduction of helmets. It is also true that some individuals were not in favour of the introduction of a generally available steel protective helmet. They were able to point out that the numbers of soldiers hospitalised due to head wounds increased once steel helmets had been introduced. It was pointed out to those individuals that many of the wounded soldiers, without the benefit of steel helmets, would not have survived their wounds.

The total of battlefield dead is truly sobering. The estimated figures are as follows:

ALLIES

Russia	1,700,000
France	1,384,000
Great Britain	743,000
Italy	615,000
Romania	335,000
Canada	60,000
Australia	59,000
India	49,000
United States of America	48,000
Serbia	45,000
Belgium	44,000
New Zealand	16,000
South Africa	8,000
Portugal	7,000
Greece	5,000
Montenegro	3,000

CENTRAL POWERS

Germany	1,800,000
Austro-Hungary	1,290,000
Turkey	325,000
Bulgaria	90,000

These figures are necessarily estimates. It is presumed that the casualties suffered by non-nation states, such as Poland, Czechoslovakia and Finland, are included in the above.

It is not possible to draw from the above figures conclusions as to the benefit derived from the use of steel helmets by any individual country. What is certain from the information available is that the casualties for war dead would be considerably larger if not for the steel helmet. The Great War represented a watershed. During that terrible conflict, protective helmets came to be regarded as an essential part of the soldier's equipment and that view has in no way changed in the intervening years.

It should be noted that captions for Imperial War Museum photographs are the actual caption given to that photograph at the time. Where we have felt it appropriate to add to the caption, our comments begin with the word "note".

Finally we mention in passing that measurements and specifications given are as precise as possible in respect of the helmet(s) examined or measured. However, undoubtedly helmets will be found that differ due to manufacturing variation and materials or period of manufacture.

AUSTRO-HUNGARY

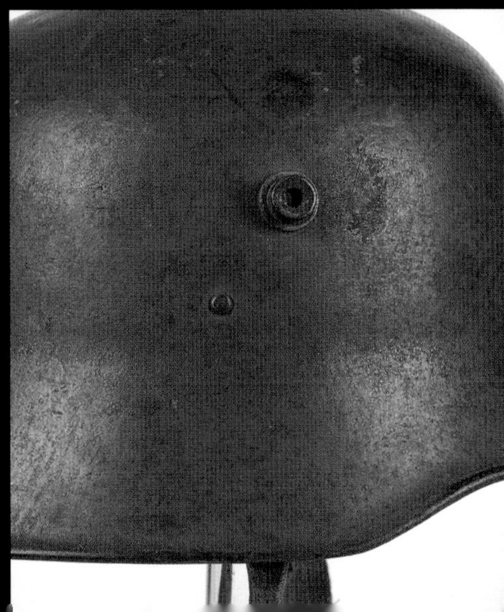

CHAPTER ONE

AUSTRO-HUNGARY

Following the war with Prussia in 1866, the Austro-Hungarian monarchy, as it existed in 1914, was founded on the 21st December 1867. On that date Franz Joseph was made Emperor of Austria and King of Hungary. Austro-Hungary and Germany were linked to one another, both by formal treaty and sentimental ties. When the Great War broke out, the two empires inevitably sided with one another.

When general mobilisation was ordered on the 31st July 1914, the Austro-Hungarian army numbered some 3.35 million men, including reservists. Of that number, actual potential front line soldiers numbered certainly no more than 2 million men, and probably rather fewer. However, the peacetime army numbered only some 450,000 men. Upon mobilisation, severe shortages of equipment quickly developed. Furthermore, although attempts to modernise the army had been made, leading to the publication of the 1907/10 Uniform Regulations, the equipment in use in 1914 was essentially 19th century in design. In common with all other combatant nations, there was no generally available protective helmet.

The headwear in use in 1914 was, firstly, the infantry shako. This was primarily reserved for ceremonial occasions. Jäger troops wore the jägerhut. This was nothing more than a stiffened felt hat and again, it was largely reserved for ceremonial purposes. For use in the field, officers and men wore the field cap or, where appropriate, the fez. The gendarmerie (military police) wore a spiked helmet. These were made of leather (Model 1899) or cork (Model 1902) or waxed linen (wartime production). Troops serving in hot climates wore a pith helmet made of cork or woven straw and covered with a khaki-coloured cloth. Mounted troops (uhlans) wore the shapka or the dragoon (hussars) helmet or the shako. The artillery also wore the shako. Where appropriate, a cloth cover could be worn with the helmet, but no helmet offered any real protection. Most early war photographs of Austro-Hungarian troops show them wearing nothing more protective than a field cap made of cloth.

From the start, there were high levels of casualties suffering from head wounds. Matters were made worse when, on the 23rd May 1915, Italy entered the war on the side of the Allies forming a further front, from the Swiss frontier to the Adriatic. This front measured several hundred kilometres and much of it crossed mountainous areas. Casualties from rock splinters, as suffered by the German forces in the Vosges Mountains, made the need for a protective helmet urgent.

The introduction of a satisfactory protective helmet for the Austro-Hungarian troops took longer than with almost any other combatant nation. This is partly explained by the severe shortages of materials and equipment already referred to. The task of providing suitable equipment and clothing to the large numbers of soldiers coming into the army was vast. Matters were made worse by the fact that the pike grey uniforms in use in 1914 were deemed unsuitable for the type of warfare that developed. In this respect, on the 16th August 1915 the Emperor decreed that fieldgrey (grey-green) was to be the standard uniform colour. This decree also applied to the cavalry. Prior to that date the cavalry had retained their traditional, brightly coloured uniforms. Substitute materials had to be developed as shortages worsened. Concentration on these problems diminished the resources available to devote to design, and put into production, a suitable protective helmet.

Additionally, early experiments and proposals in respect of protective helmets seem, to some extent at least, to have concentrated on the development of a helmet that was proof against direct infantry fire. The Technisches Militaer Kommitee (TMK) was responsible for considering the various proposals for a new helmet. In September 1916 a report was made to the Army Supreme Command establishing for the first time that protection against direct fire should not be the main priority. Rather, protection from splinters and shrapnel should be the objective. After all, a detachable brow plate could protect from direct fire.

Some of the solutions considered early in the war were as follows:

1. The TMK made an enquiry and obtained five of the armoured face masks from the German company of Becker A.G.

Steelworks at Willich/Krefeld. These masks were tested and found satisfactory save that the weight was 5.8kg. This was too heavy for general use.

2. Bohler & Co. of Kapfenberg submitted four basic helmet projects totalling in all twenty-two models. They included models based on the French Adrian Model 1915 and another based on the Austrian Dragoon helmet. These were submitted in varying thicknesses of steel.

3. The Phoenix Steelworks Joh. E. Bleckman in Murzzuschlag submitted seven helmet models. The first of these was a steel cap to be worn under the field cap in the same way as the French "Cerveliere" (see under France). Although some of the helmets submitted were field tested, none were found to be satisfactory.

4. A moulded steel plate to be mounted on the Dragoon helmet was developed. In this respect, it was necessary to remove the helmet crest before the plate could be fitted. The plate was approximately 2mm thick, and thus reasonably effective. However, the design was not widely adopted as no protection was provided to the rear of the head.

5. A simplified version of the German helmet was considered by the TMK and production of 10,000 helmets was suggested, together with the conversion of 35,000 Dragoon helmets. This order was subsequently greatly reduced. However, no steelworks could start production as the necessary machinery was not available.

6. In October 1916 the Krupp/Berndorf Company submitted test helmets. The design was approved and an order placed for production to commence immediately. The helmet was known as the Austro-Hungarian Berndorfer and is referred to in detail later.

In the circumstances there was delay and the lack of suitable helmets was an urgent problem. In this respect, some troops had taken to wearing captured Italian helmets. Indeed, in April 1917 an order was issued that troops should collect enemy helmets when out patrolling.

In the meantime, the lack of helmets was, to some extent, alleviated by the supply of helmets from Germany. Despite the fact that the Germans were short of helmets at that time, they supplied to their allies 15,000 in November and 35,000 in December 1916. These supplies continued in varying quantities in the following months. Germany also supplied manufacturing equipment and Austro-Hungary accepted the German Model 1916 helmet, with some modification, as their standard helmet. It became known as the Austro-Hungarian Model 1917. Later, the Model 1918 version of this helmet was introduced.

Thus, the main helmets used by the Austro-Hungarians in the Great War were German Model 1916 and Model 1917 supplied by the Germans, the Austro-Hungarian Models 1917 and 1918 (sometimes referred to as the German Models) and the Austro-Hungarian Berndorfer helmet (sometimes referred to as the Domestic Model). Some authors have referred to the Austro-Hungarian Model 1918 as the Hungarian Model 1917.

The German Model 1916/17 Helmets

These were the standard German helmets. It is not known precisely how many were supplied, but the total to January 1918 is said to be approximately 416,000. It is possible the total number supplied exceeds that figure. That figure should be seen in the context that Austro-Hungary had, by January 1918, produced 534,013 of the German Model and 139,968 of the Berndorfer helmet. It is not known whether or not German supplies included brow plates and German Model 1918 helmets.

It seems likely that the helmets supplied by Germany were fitted with chinstraps. However, an Austro-Hungarian woven chinstrap was, in some cases, fitted. It may be that the Austrian chinstraps were issued to replace German chinstraps that had been lost or broken. It is worth mentioning that there are two helmets shown in the photographic section that combine features of both German and Austro-Hungarian helmets. Each has a standard German helmet shell, one marked E.T.66 and the other Si.66. The former has Austro-Hungarian Model 1917 chinstrap attachments, chinstrap and liner. The latter has an Austro-Hungarian Model 1917 liner but is fitted with German chinstrap attachments and the Austro-Hungarian chinstrap designed to fit the German helmet. Both helmets are painted in Austro-Hungarian brown and have a triangular ink-stamp close to the right-hand chinstrap attachment. The stamps read R.St.U.O. CILLI. Nothing certain is presently known about these helmets, but it does seem probable that they were supplied by the Germans as helmet shells only, to be finished by the addition of the various fittings. Such a scenario is possible in the circumstances where Austro-Hungary was not yet able to produce helmet shells, or at least an adequate quantity of helmet shells, but was able to manufacture the fittings and finish the helmets by fitting the chinstrap attachments, chinstraps, liners and painting.

In any event, the Austro-Hungarian chinstrap designed to fit the German helmets is rare today. The chinstrap had two buckles, the one being a clamp-type buckle and the other a sliding buckle without prong. The attachment fittings at each end of the chinstrap were very different to those used on the German helmet. Stamped from sheet steel with folded sides, they were rectangular with a square end, instead of the German rounded end. They measured 44mm long and 26mm wide. One end had a circular hole with a triangular slot to fit onto the helmet, whilst the other end had a rectangular slot through which the chinstrap passed. To the inside of the clamp-type buckle a piece of leather, designed to protect the wearer's chin, was attached by stitching.

It appears that standard Austro-Hungarian helmets were not available at the Front until May 1917. The only helmets available prior to that, apart from the limited number of experimental or quasi-experimental helmets mentioned above, were either those supplied by Germany or captured helmets.

The Austro-Hungarian Model 1917 Helmet (German Model)

Design, Introduction and Production

This is the Austro-Hungarian version of the German helmet. The helmet shell and ventilation lugs are identical. A comparison of the two models of the same size shows that the measurements from front to rear and from side to side are almost identical. They were manufactured using the same method of marking with the manufacturer's initials or trademark and size stamped in the interior left-hand side near the chinstrap attachment.

Perhaps the best way to study this helmet is to read what is said about the Model 1916 and 1917 helmets in the German section, and then to note the following differences:

• The Austro-Hungarian helmets were painted brown. The paint used varied from a light brown to a rich, earth brown. Camouflage finishes, after the German style, were sometimes used.

• The Austro-Hungarian helmets were fitted with a woven chinstrap, attached permanently to the helmet by stitching. The Austro-Hungarian version was of two-piece construction. One part of the strap had a zinc-plated buckle at the end. The buckle had a roller and prong. The other part had six holes in it to allow for adjustment of the length. The holes were reinforced with metal eyelets. The ends of the chinstrap, where the material might fray, were covered with light brown leather held in place by stitching. The chinstraps were 20mm wide. That part with the buckle at one end was approximately 210mm long and the other part was 170mm long.

• The chinstrap lugs on the Austro-Hungarian helmet were quite different. They consisted of rectangular wire loops to which the chinstrap was stitched. The external measurements of the rectangular wire loops were 28mm by 12mm and the wire used was 3mm thick. The rectangular wire loops were fixed permanently to the helmet by means of flat pieces of metal which were rectangular with stepped, rounded ends. Each piece of metal was passed through the wire loop and folded back so that the ends met, thus securing the loop. Each metal fitting was then bent at an angle so that, once fitted, the wire loop would be away from the side of the helmet. The chinstrap attachment was then fixed to the interior helmet apron by means of a single, round-headed rivet. The rivet was inserted from the exterior of the helmet.

• The height of the chinstrap attachment above the bottom of the helmet apron varied greatly on the two helmets, the Austro-Hungarian attachments being much higher. The base of the chinstrap attachment rivet on the German helmet (Size 64) was about 22mm from the bottom of the helmet, whilst on the Austro-Hungarian helmet it was 47mm above the base. It is thus possible, when studying contemporary photographs, to determine whether any given helmet is German or Austro-Hungarian in origin.

• The Austro-Hungarian helmet was not made with a leather liner band, as used on the German Model 1916.

• The metal liner band on the Austro-Hungarian helmet, when compared with the German Models 1917 and 1918, was somewhat different. On the German helmets there was a rim pressed in the liner band near the lower edge. This was absent on the Austro-Hungarian helmets. On the German helmets there were 32 - 34 small holes punched in the liner band beneath where the liner pads were fitted. On the Austro-Hungarian helmet there were similar holes, although there were only 18 and they were larger. Additionally, the liner band joint was sometimes secured by two small rivets, a feature not found on the German helmets.

• The liner pads on the two helmets were similar. However, the hole at the end of each tongue used for the drawstring was reinforced with a metal eyelet on the Austro-Hungarian helmet.

• There was no marking on the interior apex of the helmet shell to indentify the batch of steel, as there was on the German helmets. Any marking on the Austro-Hungarian helmets is to the exterior of the top of the helmet shell.

• The sizing system was the same for the two helmets, except that it appears the Austro-Hungarian helmets were not made in size 60.

• On some Austro-Hungarian helmets an acceptance or delivery mark, stamped in ink, may be found inside the apron. These stamps usually have the year, the initials "U.O.", standing for Ubernahmeorgan (official approval), and the initials of the place of delivery such as "BND", indicating Berndorf.

The Manufacturers

The following is a list of the known manufacturers of the Austro-Hungarian helmets with, where known, their codes or trademark:

- Arthur Krupp AG, Berndorf (Bear)
- A. Western, Cilli. (AW)
- Bleckmann & Poldihutte, Klando, Bohemen.
- Bruder Lapp, Rottenmann u. Warcholowsky.
- Bruder Gottlieb und Brauchbar, Brunn. (BGB)
- C. A. Scholtz, Mateocz (CAS)
- Eissler & Cie., Wien.
- Gebruder Bohler & Cie, Kapfenberg, Styria. (GB)
- Manfred Weiss, Csepel, Budapest.
- Nadrag, Transylvania.
- Resicka, Transylvania.

The Austro-Hungarian Model 1918 (Hungarian Model 1917)

This helmet was very similar to the Austro-Hungarian Model 1917. The main difference lies in the liner system. In the Austro-Hungarian Model 1918 helmet, the liner system was similar to the German Model 1918 and to that fitted to the Berndorfer helmets. The liner headband was similar to that fitted to the Austro-Hungarian Model 1917 helmet. However, there were only approximately 15 larger, circular holes punched through the liner headband on the Austro-Hungarian Model 1918 helmet under the position occupied by the liner pads.

The chinstrap attachment was quite different on the Austro-Hungarian Model 1918 compared with the Austro-Hungarian Model 1917 and the German Model 1918. On the Austro-Hungarian Model 1918, the chinstrap attachment was fitted between the liner and helmet shell. It was held in place by means of the rivet on each side of the helmet serving to hold the liner headband in place. However, the chinstrap attachment was not attached to the helmet shell or the liner headband, being merely kept in place by the rivet passing through a hole in the chinstrap attachment. The chinstrap attachment was, to some extent, moveable and the angle of the chinstrap could thus be slightly adjusted.

The chinstrap attachment had a square base measuring, at the bottom, 35mm. Into the metal of the attachment was cut a rectangular slot to attach the chinstrap. The slot measured 23mm long by 7mm deep.

The chinstrap was made of woven material and of two-piece construction. On one side it had a buckle, without prong, for adjustment. One end of the chinstrap was attached to the buckle permanently. The other end of that part of the chinstrap passed through the chinstrap attachment on the helmet and then through the buckle. The other part of the chinstrap was attached permanently by stitching to the helmet. At the free end it had a locking buckle. To secure the chinstrap, the free end was passed through the locking buckle, which was then secured. The locking buckle at the base measured 25mm. A similar chinstrap arrangement was used on some of the Berndorfer helmets.

The liner headband had, on each side, a circular hole with a slot extending towards the rear of the helmet. The head of the rivet serving to hold the chinstrap attachment in place was passed through the hole in the liner headband on each side of the helmet. The liner headband was then pushed towards the front of the helmet, engaging the head of the rivet in the slot in the liner headband. The liner headband was then secured by means of a split-pin at the rear.

The split-pin used at the rear to secure the liner headband had a head measuring approximately 10mm wide and 4mm deep. This was intended, as on the Austro-Hungarian Model 1917 and the German helmets, to keep the strap securing the brow plate in place.

The rivets on either side of the helmet securing the liner headband were quite different from those on other helmets. The rivet head, on the exterior of the helmet, had a flat head and rounded sides and measured 6mm wide and 1.5mm deep. The rivets were 75mm above the base of the helmet, thus making it easy to distinguish this helmet externally from all other models.

It appears all the Austro-Hungarian Model 1918 helmets were made by Arthur Krupp A.G. Berndorf. The helmets were stamped, in the centre of the rear apron, with the trademark of that company, namely a bear standing on all fours and facing to the left, the helmet size being stamped immediately beneath

The Austro-Hungarian Berndorfer Helmet

Design, Introduction and Production

This helmet, described variously as the "Domestic Model" or "National Model", was the first steel helmet to be produced by the Austro-Hungary during the Great War.

The manufacturer was Arthur Krupp A.G. Berndorf. Following acceptance of the helmet in October 1916, the first bulk delivery of 8,000 helmets took place in May 1917. Production continued until November 1917 when the last 1,200 helmets were delivered. The total production is put at 139,968.

The Berndorfer helmet was not dissimilar to the German helmet in concept. Common features were the horizontal brim, a flared apron to the sides and rear and a liner secured at three fixing points. Having said that, there were numerous differences between the helmets.

The Berndorfer helmet weighed about 1.5kg, depending on the size. Approximately 6mm of the rim was folded inwards to provide a smooth finish. However, it should be mentioned that at least one example exists where the brim was folded outwards. The measurement of the size 66 helmet from front to rear of the apron was 300mm and from side to side 260mm. Thus, the Berndorfer helmets were shorter lengthways but wider sideways than the German helmets.

On about 10% of the helmets there was a small, two-pronged hook fitted to the rear. This fitting was kept in place by the liner headband screw and was designed to secure the brow-plate strap.

The Ventilation System

Ventilation was provided by four, 4mm holes in the apex of the helmet. The four holes were grouped in a square, with a bump protruding outwards in the centre. To this bump was riveted a 26.5mm wide circular covering disc. The disc stood proud of the four ventilation holes, allowing air to pass. Advice was issued to the troops that it was appropriate, to prevent excessive ventilation in stormy weather, to wind a piece of twine or string around the securing rivet beneath the circular cover, thus covering the four holes. Soldiers were also advised to wear some form of soft head-covering beneath the helmet in cold weather.

The Liner System

There were several different liner systems. The most common was a design similar to the German Model 1918 liner. It consisted of a metal liner headband approximately 25mm wide. It was pierced around the entire circumference with circular holes. On one helmet examined, the holes were 10mm wide and spaced at somewhat irregular intervals, giving a "hand-made" appearance. This is thought to be a helmet of early manufacture. On other helmets examined the holes were evenly spaced and 7mm wide. Around the lower edge of the liner headband there was a row of small holes punched in the metal. On the early helmet examined, the small holes were punched around the entire circumference, although there was a gap in the holes where the liner was attached at the rear of the helmet. On the later helmets the holes only appear under the liner pads. The purpose of the small holes is unclear since they were not used to secure the liner pads. There may have been an intention, not adopted, to use the small holes for attaching the liner pads by stitching.

There were three liner pads. They were made of brown leather, or a substitute material probably used on later production helmets because of a shortage of leather. The liner pads were very similar to those on the German helmets. They had pockets to take the liner pad cushions. The standard material intended to be used to make the pockets and to cover the liner pad cushions was cream-coloured linen. However, in reality, many different types of material were used, some of it quite colourful. One distinguishing feature is that the holes at the end of the liner pad tongues, designed to take the drawstring, have a reinforcing metal eyelet. Where leather substitute material was used, the holes at the ends of the liner pad tongues were reinforced with a circular piece of material placed on the underside.

The liner headband was held in place by three screws, one on each side and the other at the rear centre. Those on the side also served to secure the chinstrap attachments. The screw at the rear served to secure together the ends of the liner headband, although welding was occasionally also used. The lower edge of the liner headband was rolled and the three liner pads were held in place by crimping under the rolled edge. Some helmets had small, washer-like nuts on the interior into which the screws were fixed.

The other generally made liner consisted of a felt headband, to which the liner pads were attached by stitching. The method of attachment to the helmet shell was the same as with the liner system described above, save that the securing screws each had a washer or washers on the interior of the liner headband into which the screw could be tightened.

A third type of liner used a metal liner band. Attached to this were four 20mm thick felt cushions. These were sewn onto the metal liner headband using hemp twine. The liner was made of leatherette material to which was stitched a linen upper part. This upper part, or crown, could be adjusted by means of a drawstring.

Advice was given to the troops as to the adjustment of the drawstring on the various liner systems. The drawstring was to be adjusted so as to distribute the weight of the helmet across the whole head, rather than the weight being taken on the liner headband.

The Chinstrap and Chinstrap Attachment

The chinstraps were generally of woven material although leather chinstraps have been observed. The woven chinstraps were the same as used on the Austro-Hungarian Model 1917 and 1918 helmets, fitted with either a buckle with prong or a clamp buckle, and secured to the helmet by stitching. The more usual type of chinstrap fitted to the Berndorfer helmets appears to have been the type with a clamp buckle. Where fraying of the chinstrap might occur, leather reinforcing was employed. The attachment to the buckle was either by means of the woven chinstrap passing through the buckle and then being secured by stitching. Alternatively, a piece of leather was passed through the buckle and then each end of the leather was stitched to the chinstrap.

On what is believed to be an early manufacture helmet, each chinstrap attachment was made of two parts. The first was a piece of heavy wire, to which the chinstrap was attached. The other part was a bracket. One end of this bracket was wrapped around the piece of wire, leaving a space in the middle to take the chinstrap. The other end passed between the helmet shell and the liner headband and was wrapped around the top of the liner headband. The screw securing the liner headband passed through the helmet shell, the chinstrap attachment and the liner headband. The chinstrap attachment was approximately 34mm wide. However, variations exist and one helmet examined had chinstrap attachments 34mm wide on one side and 38mm wide on the other. On the later manufacture helmets examined, the piece of wire is omitted, the end of the chinstrap attachment simply being of rolled metal without the wire insert.

Paint Finishes

Early helmets appear to have been finished with a light brown paint that had an enamel-like, shiny finish. Later helmets were finished in the standard, non-reflective, brown paint used on all Austro-Hungarian helmets.

On some helmets (including other models) a circular or square patch of white or cream paint was painted onto the rear of the helmet. This was intended to assist with the identification of troops and their position from the rear.

Again, some helmets were painted with camouflage finishes as used by the Germans, although the practice does not appear to have been widespread.

Helmet Sizes

Originally the helmet liner sizes were intended to be 55 to 58cm. The size was marked on the liner and/or the rear helmet apron using an ink stamp. However, the German system of helmet shell sizing was adopted. The size (64, 66 or 68cm) was stamped in the metal inside the rear apron immediately below the Manufacturer's bear trademark.

Cloth Covers

These were used to a limited extent. It is estimated that approximately 10% of troops made use of a cloth cover.

The Berndorfer Brow Plate

A distinctive brow-plate was used with the Berndorfer helmet. It was attached to the helmet in three places, as well as being secured with a strap. It was firstly attached by means of a strip of steel to the top of the helmet. The steel strip had a large, circular end with a hole in it of sufficient size to pass over the circular metal disc covering the ventilation holes. The other end of the steel strip was permanently attached to the top centre of the brow plate. It passed through a rectangular slot, cut in the brow plate, and was then wrapped around the edge of the slot.

On each side of the brow plate was a fitting used to secure it to the helmet brim. The fittings were attached permanently, but not rigidly, to the brow plate by means of a rivet. The fittings were adjustable to allow for the difference in helmet size. Also attached to this fitting on each side were the woven securing straps. These were secured at the rear of the helmet by means of a buckle.

The Austro-Hungarians manufactured and used two types of brow plate with their German-style helmets. The year of introduction and quantities of brow plates issued are unknown. The first type was copied from the German model, the only differences being the colour of the paint finish, the markings and the fact that on the Austro-Hungarian version woven rather than leather straps were used. The example examined was constructed of metal a little thicker than the German counterpart, although this may be a one-off variation.

The other type, rarely seen today, has been described as the Linear Model. It differs in that the top and bottom are parallel and it does not curve down to cover the wearer's temples or up across the top of the helmet. Attention is drawn to the photographic section where an example is illustrated.

Variations

A helmet of German type fitted with Berndorfer ventilation and liner systems exists. However, no official documentary evidence is available. It is thought the helmet was merely an experimental model, not produced in any quantity.

In addition, Austro-Hungarian Model 1917 helmets existed with an added face visor. The visor was attached by means of a screw at each side of the helmet. The screw passed through an extension on each side of the visor and then through the hole used to secure the chinstrap. The chinstrap attachment was then secured by a small bolt on the interior of the helmet.

The visor could be raised and held in that position by means of two projections. Each of the two projections was fixed to the visor by means of two small rivets at one end. The other ends each had a circular hole. These fitted over the ventilation lugs, thus holding the visor in the raised position. Two small slots in the visor provided for vision. The visor was, of necessity, constructed of thin sheet steel and was proof against only small, relatively low velocity projectiles.

Foreign Helmets

Quantities of captured Italian helmets were used by the Austrians, sometimes following conversion. These included French Adrian Model 1915, Italian Model 1915 and Russian Sohlberg Model 1916 helmets. There is a record of a War Ministry Directive of October 1917 stating that all captured helmets were to be sent to a collecting point at Brunn-Mariaenzersdorf for processing.

The captured helmets were converted by removal of the liner and fittings, including the chinstrap attachments. Thereafter, four circular holes were made around the base of the helmet bowl, one at the front and back and one on each side. The holes at either side were used to fix Austrian Model 1917 chinstrap attachments to which a standard Austrian woven chinstrap was attached by stitching. The helmets were repainted in Austrian brown and a new Austrian liner was fitted. The liner was then fixed using dome-headed split-pins to the front and rear. The use of such helmets appears to have been limited.

The conversion of the Sohlberg helmets involved drilling new holes to serve to attach the liner and filling in the original pairs of holes used for the liner headband securing pins.

Five views of the steel plate to be mounted on the pre-war dragoon helmet, following the removal of the original crest. The interior is marked M2* inside a curved bow or bracket. (Collection of David Powers)

This example is inscribed on the left side:

Soldier's Gift
NOT TO BE SOLD
J.C. Palos U.S.A.A.S with the Ital. Army
On the right side:
For: Miss M.E. Haskell
40 Concord Ave
Cambridge
Mass.
U.S.A.

German Model 1916/17 helmet shell. Manufacturer's mark/size Si.66. Fitted with Austro-Hungarian Model 1917 liner and chinstrap with pressed steel adaptors to fit the German chinstrap lugs. The helmet is painted inside and out in Austrian brown. Note the triangular ink stamp "R.St.U.O.CILLI". (Collection of Michael Haselgrove)

Comparison of German Model 1916/17 chinstrap and the Austrian Model 1917 chinstrap with adaptors to fit the German helmet.

German Model 1916/17 helmet shell. Manufacturer's mark/size E.T.66. Fitted with Austro-Hungarian Model 1917 liner and chinstrap attachments. The helmet is painted inside and out in Austrian brown. Note the triangular ink stamp bearing the name "R.St.U.O.CILLI". (Collection of Michael Haselgrove)

Austro-Hungarian Model 1917 helmet (German Model).

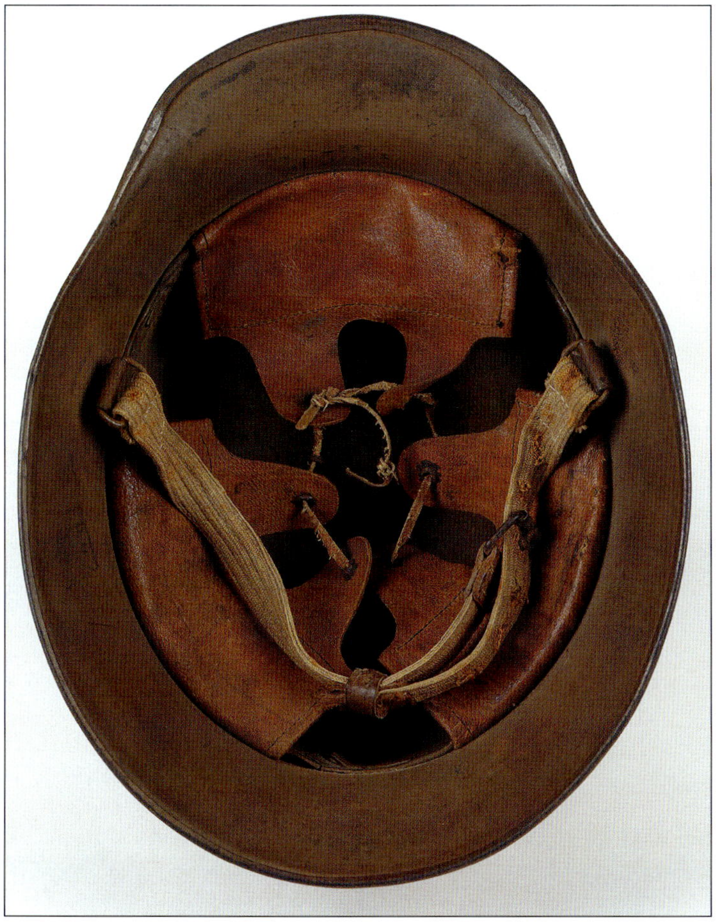

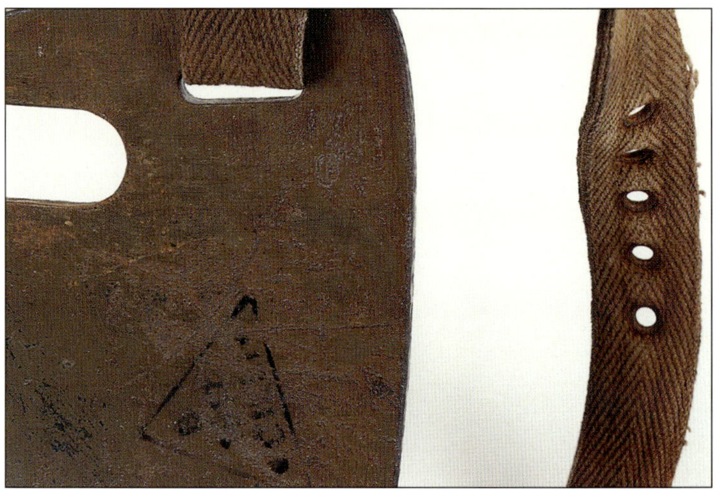

Austro-Hungarian brow plate to fit the German style helmets. Note the Berndorfer manufacturer's mark and ink stamp. (Collection of Brana Radovic)

Austro-Hungarian Model 1917 brow plate with straps attached by stitching. (Collection of Ellwood von Seibold)

Austro-Hungarian Model 1917 brow plate with straps attached by rivets. (Collection of Brana Radovic)

Waxed paper oakleaves secured to left-hand ventilation lug and worn for ceremonial occasions. (Collection of Brana Radovic)

Austro-Hungarian Model 1917 helmet with German-style camouflage paint scheme. This helmet has had a shiny varnish finish applied, probably post-war. (Collection of the Imperial War Museum, London)

Austro-Hungarian Model 1918 (Hungarian Model 1917) Helmet. (Collection Marcus Cotton)

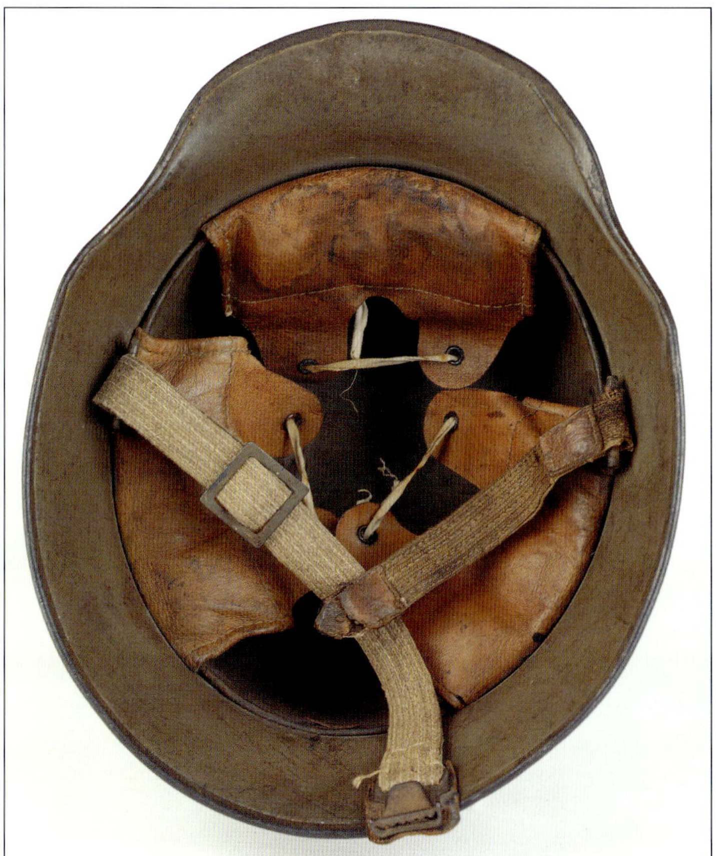

Austro-Hungarian Berndorfer Helmet. Marked in the centre rear with the bear trademark of Arthur Krupp A.G. Berndorf and size 66. The shiny, enamel-like finish is said to have been applied to early helmets. Note the identification mark applied to the rear exterior of the helmet. (Collection of Michael Haselgrove)

Austro-Hungarian Berndorfer Helmet. Note the shiny paint finish and felt liner headband. (Collection of the Imperial War Museum, London)

Austro-Hungarian Berndorfer Helmet with matt paint finish. (Collection of Marcus Cotton)

Austro-Hungarian Berndorfer Helmet with matt paint finish. Note leather substitute liner material. The helmet shell is marked to a 38cm howitzer battery. (Collection of Brana Radovic)

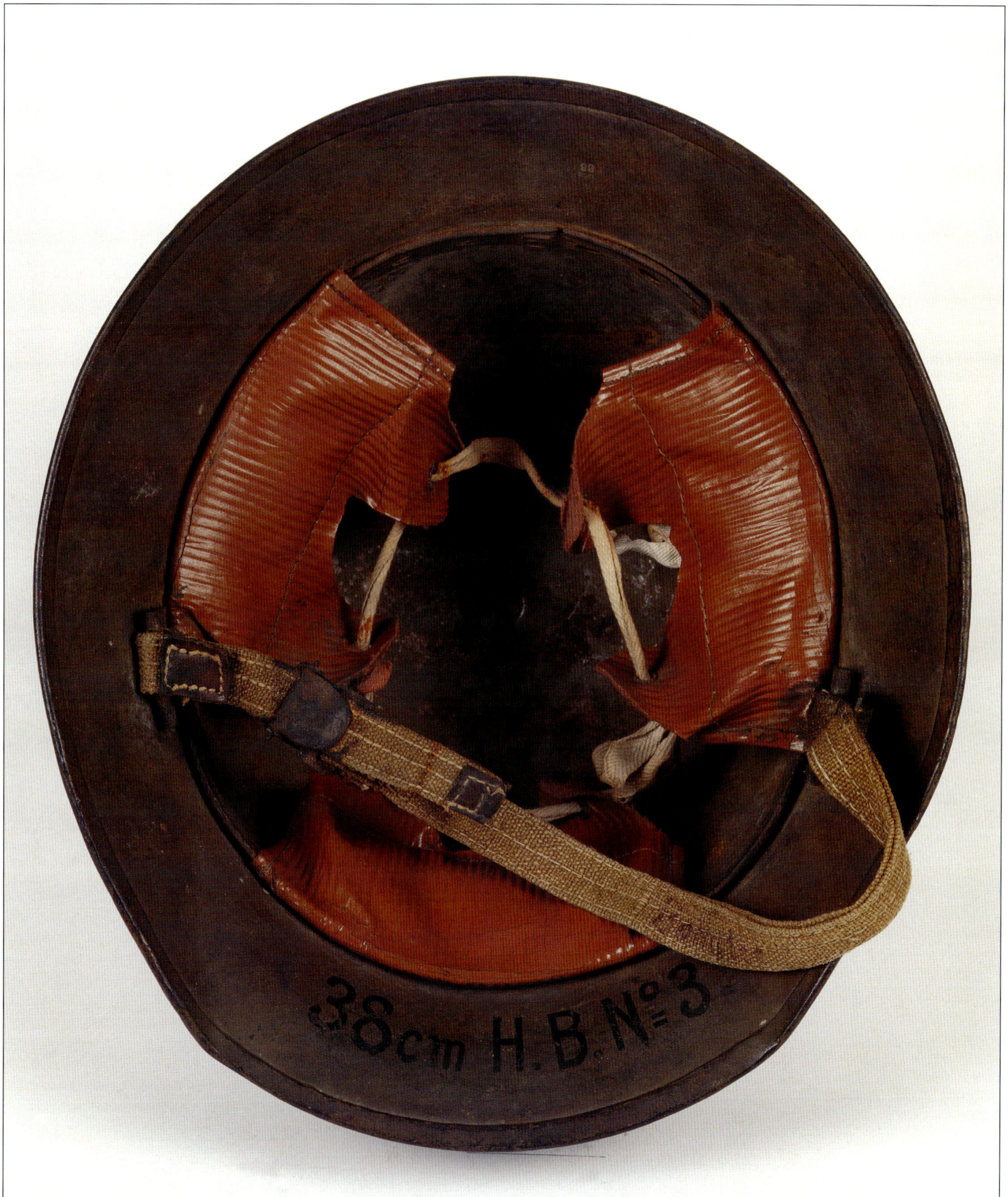
38cm H.B.№3

Austro-Hungarian Berndorfer Brow plate. (Collection of Marcus Cotton)

Bottom: Battle-damaged Austro-Hungarian Berndorfer Helmet with painted identification mark and two-pronged hook used to secure the brow plate strap. Note the hook appears to have been turned upside-down on this example.

Austro-Hungarian Berndorfer Helmet with painted camouflage finish. (Collection of David Powers)

Austro-Hungarian Berndorfer Helmet fitted with drawstring cloth cover. (Collection of Brana Radovic)

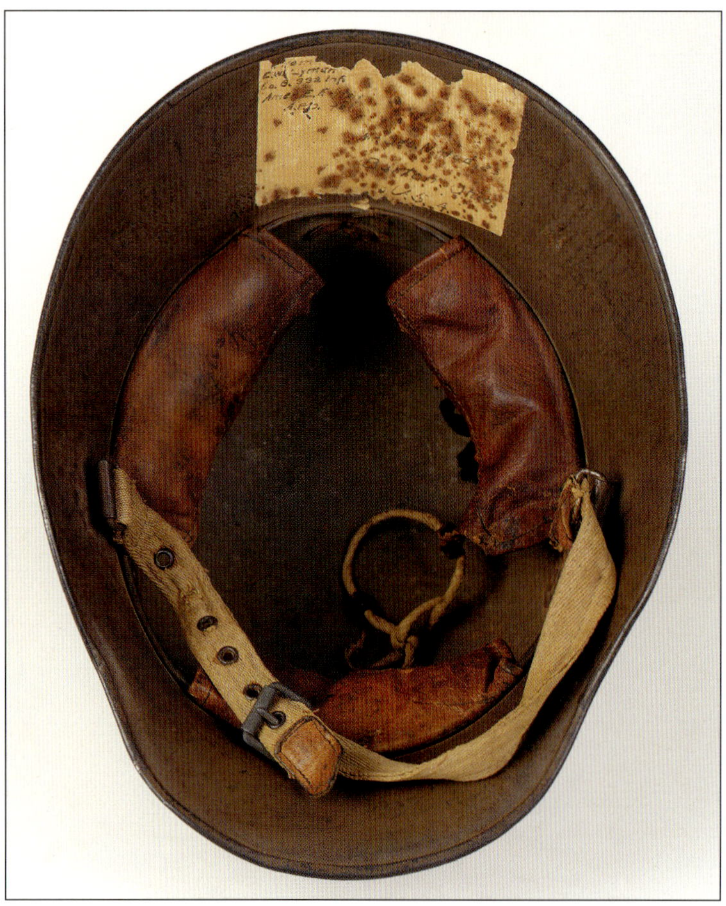

Standard Austro-Hungarian Model 1917 Helmet with post-home labels. The labels read: "From E.W.Lyman CO. G 332 INF. Amer. E. F. Italy A.P.O. 701 to MR Geo. N. Hodges, Burton, Ohio, USA" (Collection of Michael Haselgrove)

Another example of the standard Austro-Hungarian Model 1917 Helmet with post-home labels. (Collection of Brian Maederer-Military Collectibles)

French Adrian Model 1915 Helmet. Supplied to the Italians, it is without badge attachment slots. The helmet was captured and converted by the Austro-Hungarians and has a replacement liner, chinstrap attachments and chinstrap fitted. This helmet was repainted Austrian brown inside and out. Subsequently, it has been painted blue externally and to the interior brim. (Collection of Michael Haselgrove)

An Italian Model 1915 Helmet captured by the Austro-Hungarians and converted by repainting and replacement of the liner, chinstrap attachments and chinstrap. (Collection of Michael Haselgrove)

An Italian Model 1915 Helmet captured by the Austro-Hungarians. Repainted and with Austro-Hungarian chinstrap attachments fitted. Although the liner is absent, there are holes drilled to the front and rear of the helmet shell to take the split pins that would have secured the liner. (Collection of Michael Haselgrove)

On this page and overleaf is shown a Russian Sohlberg Model 1916 Helmet, captured and converted by the Austro-Hungarians. The liner and chinstrap fittings have been replaced by Austro-Hungarian Model 1917 fittings. The rivet holes that secured the original liner have been plugged. The helmet has been repainted and the leather chinstrap is of a non-standard type. The distinctive patterned cloth forming the liner pad pockets is of a type observed as used in a standard Austro-Hungarian Model 1917 Helmet. Note that one of the leather liner pads has been replaced, probably post-war. (Collection of David Acon)

Austro-Hungarian Model 1917 painted as a souvenir (Collection Robert Tredwen)

Posed photograph showing troops wearing Austro-Hungarian Berndorfer helmets.

Top: Type of Austrian prisoner. Note Austro-Hungarian Berndorfer helmet fitted with what appears to be a leather chinstrap. IWM Q26815

Bottom: Type of Austrian prisoner. Note he wears what appears to be an Austro-Hungarian Model 1918 (Hungarian Model 1917) helmet with a woven chinstrap. IWM Q25861

Soldier wearing the Austro-Hungarian Model 1917 helmet.

Top: Austrian first aid post in the Frenzella Ravine. June, 1918. Note identification paint applied to the rear of the helmets of the two soldiers at the centre of the photograph. IWM Q88465

Bottom: Kaiser and King Karl reviewing stormtroops in the South Tirol. Note that the troops are wearing Austro-Hungarian Model 1917 helmets. Note variation of paint finishes. IWM 112589

Oberleutnant Kern of Infantry Regiment No.14 after receiving his medals. IWM Q112617

BELGIUM

CHAPTER TWO

BELGIUM

Belgium occupied a strategic position prior to, and during, the Great War. Her position was, in effect, that of a strategic military route, leading in both directions, between Germany and France. Prior to the war, Belgium was neutral and much depended upon which country violated that neutrality first. In 1911, the British cautioned the French not to be the first to cross the Belgian frontier. If they did, the British would not be on the side of the French. If Germany violated Belgian neutrality first, that would bring the British into the war against Germany.

Whilst King Leopold II was on the Belgian throne the Germans were confident that the issue of Belgian neutrality could be resolved by the payment of money, the Germans relying upon the avarice of King Leopold. Even when King Albert succeeded to the throne in 1909, the Germans continued to expect Belgian resistance to their invasion to be a formality. The steadfast Belgian resistance against a numerically vastly superior force surprised not only the Germans, but the allies of the Belgians as well.

When Germany invaded Belgium on the 4th August 1914, the Belgian Field Army numbered only 117,000 men. The trained reserves were all deployed manning the fortresses that formed an important part of the Belgian defences. There was, in addition, the Belgian Civic Guard, whose helmets in the form of a top-hat feature in early war photographs.

The Belgian army in August 1914 was poorly equipped for the type of warfare that quickly developed. The question of a suitable protective helmet had been considered before the war. Numerous suggestions were put forward. In 1913 a helmet designated the Wahis-Dillens Model 1913 was adopted. Lieutenant Wahis was an aviator and horseman. Gustave Dillens was a sculptor. The design of the helmet was rooted in the nineteenth century. It was constructed of felt with metal ornamentation. This included a sculptured lion's head at the front of the crest. The helmet was totally unsuitable for the conditions of the Great War and it is perhaps fortunate that only a few units were equipped with this helmet by the time the war started. Most of the troops wore a cloth kepi.

The Belgian Adrian Model 1915 Helmet

Initially, at the start of the Great War, the Belgian army considered the adoption of an experimental French helmet. However, by October 1915 the French began delivery of the first Adrian helmets to their allies, including Belgium. Approximately 210,000 helmets had been delivered to the Belgians by the end of 1916.

The Belgian Adrian Model 1915 was identical to the French helmet save for the badge. The helmets were painted light brown. Various shades of brown and different paint finishes may be encountered. The badge was a frontal view of a lion's head in relief and was 65mm high and 54mm wide. All Belgian troops including officers wore the same helmet and badge, with no separate badges for artillery, engineers etc.

Some short time after the helmet came into service khaki cloth covers were issued. A feature found on some Belgian covers was a slot on each side of the crest to allow for ventilation.

The Belgian police also used the same helmet from April 1917. Police helmets were painted blue-grey. Subsequently, the police helmets were painted black.

Lightweight helmets, constructed of cork and canvas, exist. These were designed for use on parade and other ceremonial occasions. However, the two examples examined were made respectively in Brussels and Liege. Throughout the war those two cities were in German hands. It is therefore likely that the helmets were made after the war.

For more detail in respect of the Adrian Model 1915 helmet, the reader should turn to the section under France.

The Model 1917 Queen Elisabeth Helmet

Despite the fact that the Belgian Army received adequate supplies of helmets from the French, there remained a wish to have a distinct Belgian helmet. Queen Elisabeth of Belgium, who was concerned for the welfare of the troops, was particularly interested in the adoption of a helmet that would protect the face and eyes of the wearer. She gave support and financial assistance to the development and production of such a helmet and for this reason the helmet that was eventually adopted bears her name.

Tests were carried out on a number of helmets, including a Dunand model. None were found satisfactory and the matter was placed in the hands of Mr. John MacIntosh who was Director General of the British Trench Warfare Supply Department. However, it is Professor Weekers, an Ophthalmologist, who is credited with the design of the helmet. There is no doubt that he was influenced by the designs of the Dunand and Polack helmets and visors.

The Model 1917 Queen Elisabeth Helmet was introduced in December 1917 and saw service on the Yser Front in 1918. The helmet itself was a one-piece construction with an added visor. The visor could be worn in the raised or lowered position. On each side of the helmet, where the visor was attached, there was a rosette in the Belgian colours of red, yellow and black. It is commented that the test helmets pictured in *Helmets and Body Armor in Modern Warfare,* by Bashford Dean, do not have the rosette fitted.

The firm of Sankey of Wolverhampton in England manufactured the helmets. In his book, Bashford Dean comments that about 40,000 of the helmets were manufactured and used for special service. It is thought that, in reality, only approximately 1,000 of the helmets were made. The helmet should really be regarded as experimental only and is rare today. The helmet shells have British stampings, and were fitted with a British liner and chinstrap.

It is worthwhile considering what Bashford Dean wrote about the helmet in *Helmets and Body Armor in Modern Warfare*. He wrote:

> ***Manufacture***
> *It was evidently found difficult to reproduce so deep a model in the manganese alloy which had been used for the regular British helmet, still the present model is probably pressed in this steel. It could not, however, have been made of it, if we accept the report from Major Dunning that this helmet had been "retempered", for this would imply that it had been given a heat treatment during its manufacture, a process which would have injured a helmet if made of manganese alloy.*
>
> ***Ballistic Results***
> *Actual tests of the helmet have yielded good results, according to Mr. MacIntosh's data. In a report from the Woolwich Arsenal, it was shown to be proof at 730 foot seconds (bullet weighing 230 grains, automatic revolver) and sometimes to resist at 820 foot seconds. The visible result of a test of this helmet in Washington appears in Fig. 116; an impact of 602 foot second automatic revolver bullet failed to penetrate but produced an area of indentation which indicates strongly that the helmet is made of manganese steel. A shot in the forehead was resisted but showed again a very great indentation. The visor perforated readily at 602 foot seconds and the perforated visual areas were found to have relatively little resistance, probably not as high as 200 foot seconds.*
>
> ***Weight***
> *The weight of this helmet is twenty-eight ounces without lining, thirty-six with lining, and forty-four with visor complete.*
>
> ***Thickness***
> *The helmet is well pressed although thinned out considerably at the crown. Near the rim of the helmet, its metal measured .044 inch in thickness and at the top .035 inch. The minimum indentation noted in the above ballistic test is one half inch measured from a line connecting the sides of the crater. The thickness of the visor is .026 inch.*
>
> ***Critical Notes***
> *This Anglo-Belgian helmet is comfortable to wear and in general, considering that it is a visored helmet, its balance is good. Its lining is in the French style, that is, having a continuous leather head-covering which terminates above in a number of small lappets held together by a string. The lining is supported by a separate carrier which is attached to the shell of the casque by coiled springs, after the fashion of the Dunand model. In the matter of shape, the present helmet is criticized as being too narrow and flat to insure the maximum safety in use. Even a relatively slight shock in the temple region would be apt to produce grave injury to the wearer. In another direction, the form of the bowl of the helmet is faulty, for it is highly arched at the back where the head almost touches the helmet shell. No statistics are at hand to show in what number this helmet was introduced in the Belgian Army. (Later specimens examined are provided with lining after the English Model.)*

The authors comment that some confusion may have arisen between the Model 1917 Queen Elisabeth Helmet and the Franco-American designed helmet incorporating a Dunand visor. The liner system described by Bashford Dean, with its attachment to the helmet by coiled springs, exactly matches the liner system fitted to the Franco-American helmet. It is possible that a Model 1917 Queen Elisabeth helmet was fitted with the Franco-American liner system, but that does seem unlikely. The above test results therefore probably relate to the Franco-American helmet rather than to the Queen Elisabeth.

The Italian Farina Helmet

Photographic evidence shows that the Belgian Army made use of the Italian Farina helmet and body armour. However, it appears that the numbers used were small and the helmet was not adopted officially. For further information about these helmets, see under Italy.

Belgian Adrian Model 1915 Helmet. (Collection of Marcus Cotton)

Belgian Adrian Model 1915 Helmet. (Collection of Ellwood von Seibold)

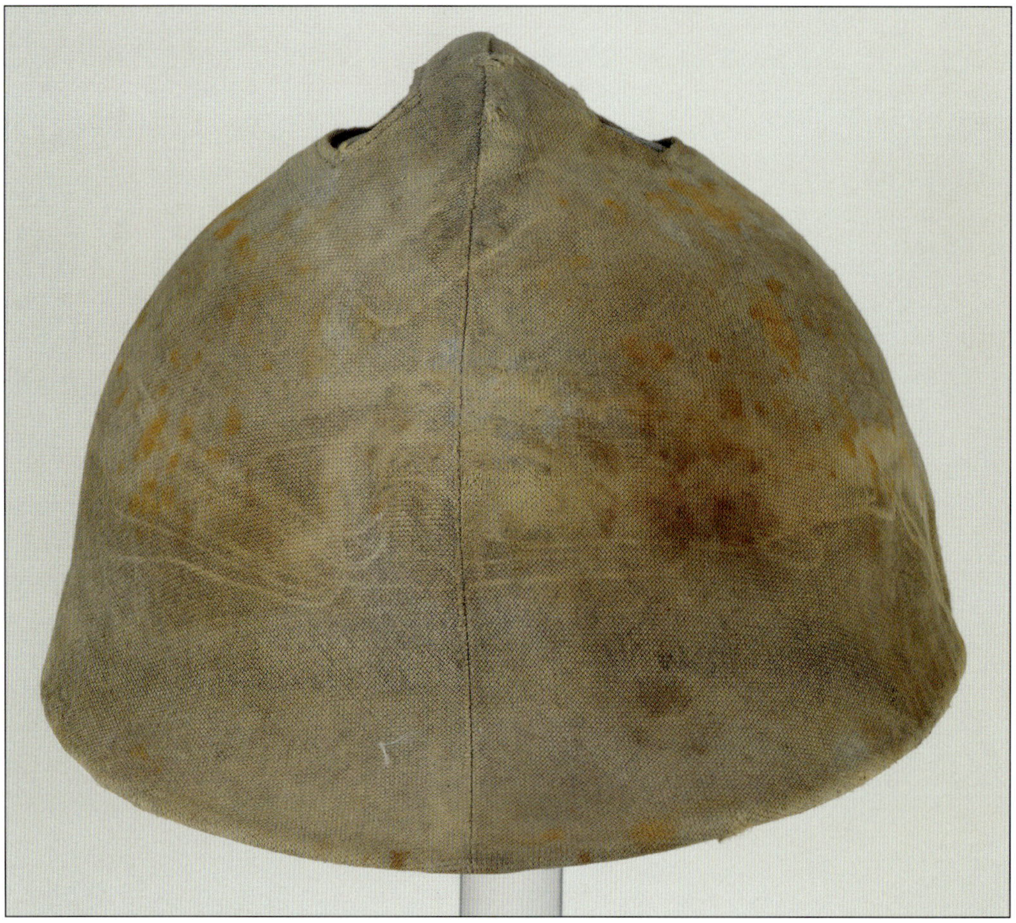

Belgian cloth cover (Collection of Brana Radovic)

Battle-damaged Belgian Adrian Model 1915 Helmet. (Collection of Michael Haselgrove)

Model 1917 Queen Elisabeth Helmet. (Collection of Imperial War Museum, London)

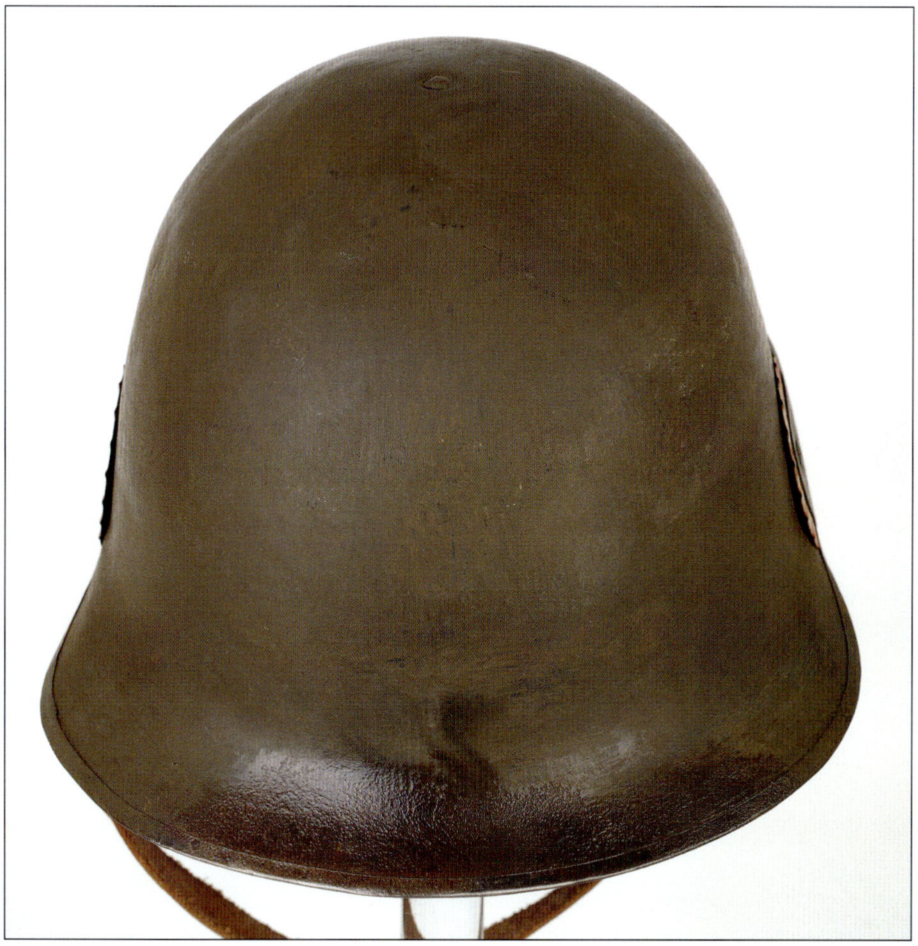

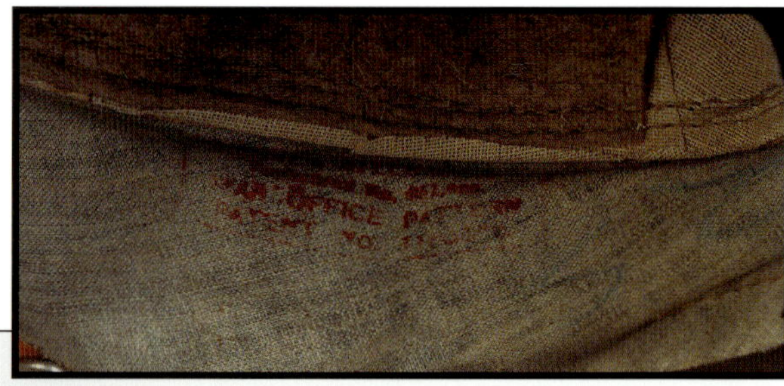

Ink-stamped British War Office mark inside liner pad.

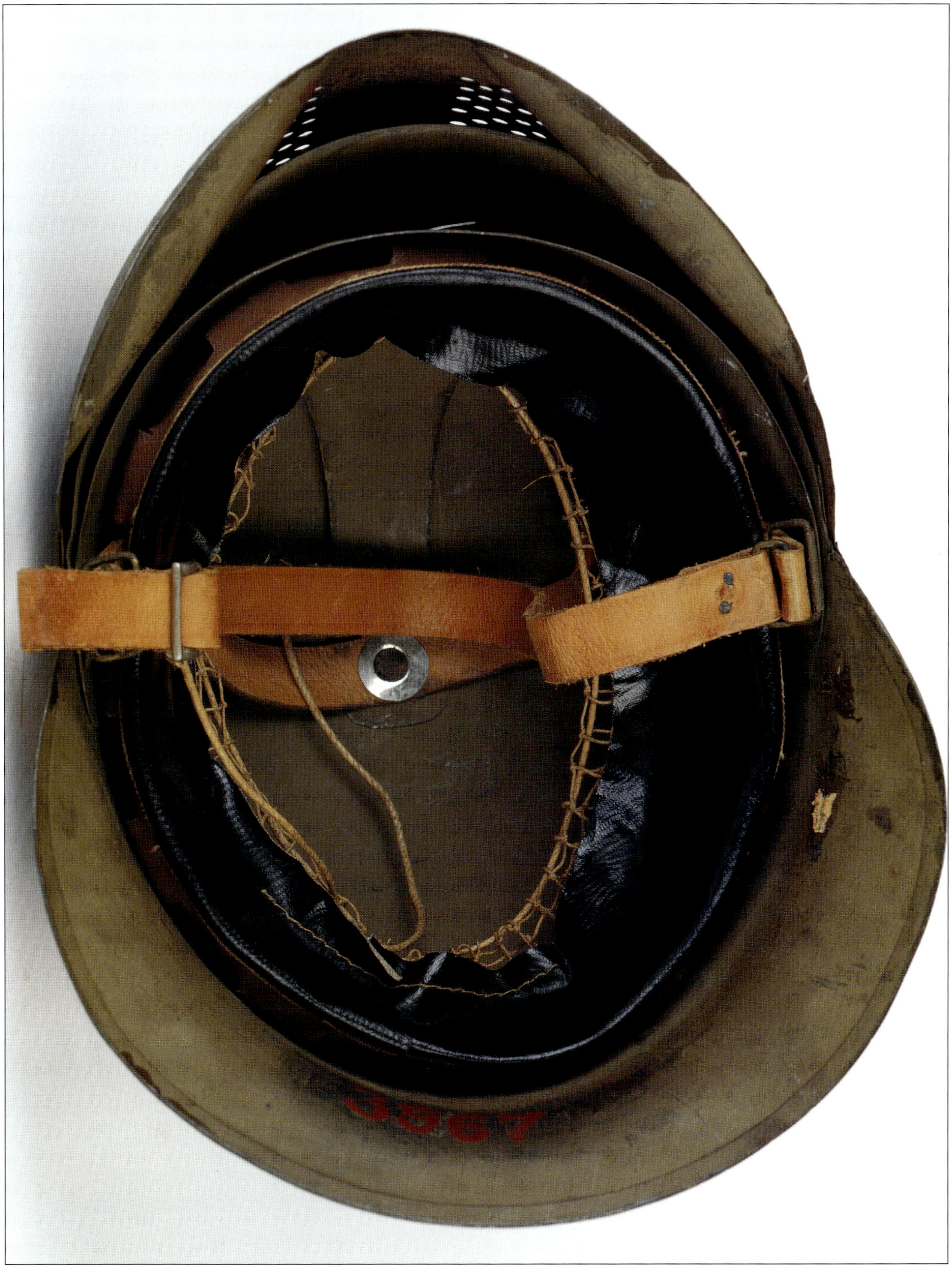
3367

Another example of the Model 1917 Queen Elisabeth Helmet. (Collection of Ellwood von Seibold)

Detail of rosette and attached French-style private purchase chinstrap. (Collection of Brana Radovic)

Studio portrait of a soldier wearing a Belgian Adrian Model 1915 helmet.

Top: The Queen of the Belgians visits the Belgian front. Note that she wears the Belgian Adrian Model 1915 Helmet. IWM Q80721

Bottom: Journalists of Calais visit the front. Note the use of Belgian Adrian Model 1915 helmets and the soldier with a cloth cover on the right. IWM HU 93200

Belgian troops cleaning their kit. IWM Q2954

■ Top: 1917. Machine Gun Section on the way to the trenches. (IWM) Belgian Official Photograph No. 55

■ Bottom: Belgian soldiers made prisoners by the Germans.

BULGARIA

CHAPTER THREE

BULGARIA

In August 1914 Bulgaria was a significant, if not dominant, power in the Balkans. However, when the war started she remained neutral. Bulgaria had been involved in the First and Second Balkan Wars in 1912 and 1913 and there was little enthusiasm for war. Additionally, it was unclear where her best interests lay. In those circumstances, in common with Greece, Italy, Portugal, Romania and Turkey, Bulgaria delayed her entry into the war.

The Allies were hopeful that Bulgaria, along with other countries, would be brought into the war on the side of the Allies by the campaign in the Dardanelles. Much depended on the success of that offensive. Bulgaria had sided with Greece and Serbia against Turkey in the First Balkan War and on the side of Turkey against Serbia and Greece in the Second Balkan War. In the circumstances, the Allies did not trust the Bulgarians to remain neutral and wished to see a Bulgarian attack against Turkey, thus clearly demonstrating where Bulgarian loyalties stood.

Bulgaria remained neutral whilst there was the prospect of success for the Allies in the Dardanelles. However, that neutrality was not as strict as it might have been. Bulgaria permitted German war material and personnel to cross her territory en route to assist the Turks at Gallipoli, where the Allies had not achieved the confidently predicted advances. They had instead become bogged down in trench warfare.

With the Allies stalled in the Dardanelles and suffering reverses elsewhere, Bulgaria considered her territorial and other interests were best served by an alliance with the Central Powers. A secret treaty was signed between Germany, Austro-Hungary and Bulgaria on the 17th July 1915. A formal treaty was concluded on the 6th September 1915.

When the Austro-German offensive against Serbia was launched on the 5th October 1915, Bulgaria decided to end her neutrality. On the 11th October 1915, she too attacked Serbia. In the days following, she declared war against the other Allies.

Serbia could not resist the onslaught, despite the fact that the Allies rushed to her aid with French troops advancing to the border between Greece and Serbia and the opening of the Salonica Front, where French and Bulgarian troops first clashed on the 21st October 1915. On that front the Bulgarians, along with troops from Germany and Austro-Hungary ultimately faced British, French, Italian, Serbian and Russian forces.

Pursuant to the terms of their treaty, Germany had agreed to supply military equipment, including helmets, to the Bulgarians and, by the end of 1917, approximately 170,000 standard German helmets had been delivered. Those helmets were designated the Bulgarian Model 1917.

At the end of the Great War, Bulgaria had to surrender to the Allies large quantities of military equipment. Her army was limited to 20,000 men. The German helmets continued in use, often with a shield comprising the national colours of white, red and green painted on the right side, until replaced by the Bulgarian Model 1936 helmet.

The Bulgarian Model 1917

The helmets supplied by the Germans were identical to the standard German helmet. The helmets, or at least some of them, were marked on the inside by ink-stamp. The stamp, when translated, stated "Sofia - Bulgarian Army - 1917".

It appears that some of the helmets were fitted with a new liner very similar to that used on the Adrian helmet, having a crown formed from six tongues. It is not known whether these liners were fitted during or after the war.

The Use of Foreign Helmets

It is possible that the Bulgarians made use of captured Serbian Adrian Model 1915-16 helmets during the Great War. Certainly the number of helmets provided by Germany was insufficient to provide every Bulgarian soldier with a helmet. In the circumstances, it is possible, as in the case of Austro-Hungary, that captured helmets were used.

In the following pages are shown photographs of what appears to have been a Serbian Adrian Model 1915-16 helmet. The likely history of the helmet is that it was captured during the Great War and was repainted with the original badge in place. The repainting has left a ridge of paint around the original badge. That badge appears to be that of the Serbian Adrian Model 1915-16 helmet, rather than the later badge adopted upon the creation of Yugoslavia. The badge was removed, following repainting, and the helmet has again been repainted and a Bulgarian Model 1936 liner has been fitted.

The above comments are, to some extent, speculation. It is, of course, possible that the helmet was captured during the Second World War.

German supplied Bulgarian Model 1917 Helmet. Note that the helmet has had its original German finish over-painted. (Collection of Michael Haselgrove)

Although the interior of the Bulgarian Model 1917 appears identical to the German Model 1917 Helmet, this example shows crude repairs/modifications that do not appear to be of German origin. In particular, both chinstrap lugs have been altered, on one side with the addition of a large washer to retain the chinstrap attachment. The chinstrap attachments are again crudely made and the chinstrap itself does not appear to be of standard German issue, being secured by rivets rather than stitching and with the sliding buckles being of thinner gauge metal than normal.

Note that the liner pads are stitched to the liner headband contrary to the usual practice.

Serbian Adrian Model 1915-16 Helmet believed to have been captured by the Bulgarians in the Great War and converted. Note the outline of the original Serbian shield. (Collection of David Acon)

CZECHOSLOVAKIA

CHAPTER FOUR

CZECHOSLOVAKIA

At the start of the Great War the three provinces which were later to form Czechoslovakia, namely Bohemia, Moravia and Slovakia, formed part of the Austro-Hungarian Empire. Soldiers from these provinces fought on the side of the Central Powers and on the side of the Allies during the Great War.

However, there was strong sentiment in favour of the formation of an independent country of Czechoslovakia. The best way of achieving that was seen by many to be to fight against the Central Powers. Thus formations of Czech soldiers, known as Czech Legions, were formed in France, Italy and Russia to fight against the Central Powers. These legions used the standard equipment of the army in which they were serving. For instance, those fighting in Italy were equipped with the standard Italian Model 1915 helmet. However, their helmets bore various badges, mostly featuring the Bohemian two-tailed lion, painted in either red or white depending on the unit of the wearer.

Czech soldiers might find themselves fighting against one another. This occurred, for instance, when General Brusilov launched the Russian offensive in Eastern Galicia on the 1st July 1917. Czech soldiers fighting alongside the Russians found themselves facing Czech troops of the Austro-Hungarian 19th (Czech) Division. They persuaded those fighting for the Austro-Hungarians to desert to the Russians.

On the 30th June 1918, Thomas Masaryk, with the approval of the United States, signed an agreement on behalf of his fellow Czechs whereby the Slovaks would be given an autonomous administration. Two weeks later in Prague, the National Council of Czechoslovakia was established. The British government, on the 9th August 1918, was the first to recognise the National Council as the present trustee of the future Czechoslovak government. Thus, if the Allies prevailed against the Central Powers, the Czechs would, by their long struggle, achieve their goal.

Czech Troops fighting with the French

From the end of August 1914, soldiers from the three provinces were accepted into the French Foreign Legion. In 1915 these soldiers, in common with other troops, were issued with standard French Adrian Model 1915 helmets. The helmets used by the Foreign Legion were painted light brown and fitted with the standard infantry badge.

In 1917 it was decided to form a Czech Army within the French Army. That army was officially formed on the 16th December 1917. Troops coming from Russia, Serbia, Romania and the United States of America, and no doubt other countries, were incorporated into this army. Prisoners of war, mostly captured on the Italian Front, also volunteered. The Czech Legion, as the army has come to be known, mostly fought in the Vosges sector.

The Czech Adrian Model 1915-1918

The helmet supplied to the Czech Legion in France was the standard French Adrian Model 1915 helmet. The helmets were painted blue. However, they had a special badge. This consisted of a main shield featuring the Bohemian lion. Above that was another shield with the double Slovak cross above an illustration of the three mountain ranges of Slovakia. To the left was the emblem of Moravia and, to the right, that of Silesia. The badge was 73mm high and 68mm wide and was attached to the helmet in the same way as the standard French badges. In all other respects, the Czech Adrian was identical to the standard French Adrian Model 1915. For further information, please refer to the section under France.

Czech Adrian Model 1915-18. (Collection of Ellwood von Seibold)

Czech Adrian Model 1915-18. (Collection of Marcus Cotton)

Czech Adrian Model 1915-18. (Collection of Polish War Museum, Warsaw)

Czech Adrian Model 1915-18 Artillery Special. For further information see under France. (Collection of Brana Radovic)

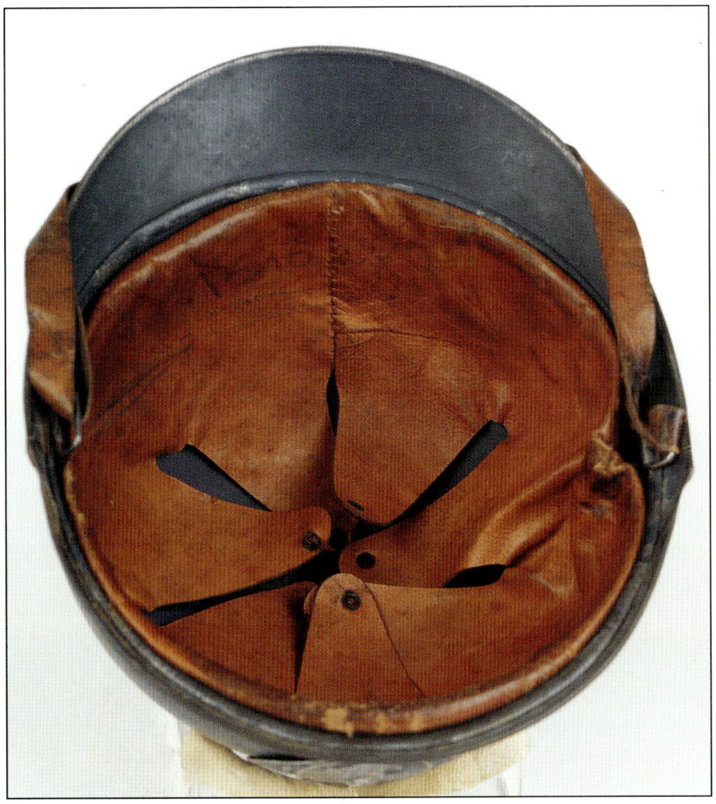

Troops of the Czech Legion wearing what appear to be Italian Model 1915 helmets. Image supplied by Nik Cornish at Stavka Military Image Research

FINLAND

CHAPTER FIVE

FINLAND

Finland became part of the Russian empire in 1808 when the Swedes were driven out. The Finnish Army was absorbed into the Russian Army and Finnish nationals serving in the armed forces were equipped in the same way as the Russian Army.

During the Great War, not all Finns served with the Russians. For instance, on the 8th August 1915 the Kaiser authorised the formation of a Battalion of 2,000 troops recruited entirely from Finland. These troops would be used to fight against the Russians on the Eastern Front. The Finnish recruits were smuggled out of Finland since, at the time, the country was under the firm control of the Russians.

Following the Russian Revolution, on the 6th December 1917 Finland declared her independence from Russia. In the heart of Russia, although there was anarchy and disorganisation, the revolutionaries were supreme. However, on the outer fringes those loyal to the old regime gathered together. The Revolutionaries in Russia were content to allow, or at least not to oppose, Finnish independence. However, they were not prepared to allow the old ruling class to resume control. In the circumstances in Finland a bitter civil war developed between forces loyal to the old ruling class, or "White Finns", and those loyal to the revolution, the "Red Finns".

The "White Finns" needed external support and for this they turned to Germany. This suited the Germans very well. On the 3rd March 1918 the Treaty of Brest-Litovsk had been signed. Russia had given up all claims to her Baltic possessions, Poland, White Russia (now Belarus), Finland, Besserabia, the Ukraine and the Caucasus. Germany sought to spread her influence inside Russia, ostensibly to enforce the terms of the Treaty. Germany had long held the wish to establish a U-boat base on the Russian Murman coast close to the Russian border with Finland and not far from the ports of Murmansk and Archangel. The naval base would give her U-boats easy access to the North Sea and the port could be used to transport oil, foodstuffs, minerals etc. from Russia, thus avoiding the British naval blockade.

In the circumstances, Germany signed a political and commercial treaty with Finland on the 5th March 1918. The arrangement was that Germany would offer support against the Bolsheviks. In return the Germans would obtain the use of an outlet on the Murman coast in territory that Finland hoped to acquire from Russia. Germany landed 20,000 troops at Hango on the Gulf of Finland. They and the Finnish forces quickly took the capital and, on the 16th May 1918, more than a century of Russian domination and six months of revolutionary control came to an end. The troops landed included those previously recruited by Germany from Finland. They were hailed as liberators of the Finnish people.

The allies had established a base at Murmansk on the Arctic Ocean and another at Archangel on the White Sea. These bases had been designed to assist with the supply of the Russians prior to the Revolution. Although on Russian territory, they were now threatened by the co-operation between Finland and Germany and by the Russian Bolsheviks. Archangel, in particular, was of great importance since, by the end of 1917, some three to four million tons of munitions had been landed and were stored there, as well as large quantities of other war materiel. The British sent a naval expedition to Murmansk where they initially established a working coalition with the Bolsheviks. However, this did not last, possibly due to German pressure on the Russians, and the Bolsheviks were driven out. Later, Archangel was occupied.

With the decline of Germany, German troops were withdrawn from Finland and Finland ceased to pursue a path of territorial gain. The worst threat to Murmansk and Archangel had been removed, although the Allies continued for a time to oppose the Bolshevik forces.

The Finnish Sohlberg Model 1915-1917 Helmet

These helmets were developed and produced in the Russian-controlled territory of Finland between 1915 and 1917 with the intention that that they would become the new Russian helmet.

The factories of G.W. Sohlberg and V.W. Holmberg produced the helmets. Sohlberg helmets were also manufactured in Russia.

There is some confusion as to why the helmets were referred to as the Model 1915-17. The probable reason for this was that the order for the helmets was delayed. It was not actually placed by the Tsarist government until 1917.

The helmet was externally very similar in appearance to the French Adrian Model 1915 helmet. The main differences were that the Sohlberg helmets were pressed from a single piece of 1mm steel with a plain edge and a circular ventilation hole covered not by a crest but rather by a triangular, dome-shaped piece of steel. Additionally, the Sohlberg helmets were painted dark green. Painting appears to have been by dipping rather than spraying.

Several Sohlberg helmets have been examined, one being thought to be of early manufacture, with the others being of later production. Although the helmets were very similar, there were various differences. It is not possible to say whether the helmets represented different types, or modifications of the same type, or were simply slight variations produced by different manufacturers.

The Sohlberg helmet examined, believed to be of early manufacture, measured 290mm from front to rear and 214mm wide. Other examples, believed to be of later manufacture, measured 284mm from front to rear and 215mm wide. The widths of each of the three sides of the metal ventilation hole cover measured 53mm on the early helmet and 52mm on the later helmets. On all Finnish helmets each of the three points of the triangular cover was fixed to the helmet shell by a single rivet, the heads of which were dome-shaped and measured between 3mm and 4mm. A small semi-circular piece of metal was cut out from each side of the triangular-shaped ventilation hole cover to allow the passage of air.

The chinstrap attachment loops were made from 2mm thick wire on all helmets examined. The wire was formed into rectangular loops measuring 21mm long and 11mm wide on the early helmet and 20mm long and 12mm wide on the later helmets, in all cases measured from the outside. Each chinstrap loop was attached to the helmet shell by means of a rectangular piece of steel measuring 11mm wide and 15mm deep. These were fixed on each side to the helmet brim by means of two rivets of the same size as those used to secure the ventilation hole cover.

The chinstrap on the early helmet was of brown leather 15mm wide and 1.5mm thick. On the later helmets the chinstraps were again of brown leather but measured 17mm wide and 2mm thick. In all cases, one end of the chinstrap was attached to a sliding buckle measuring 22mm square and the other end passed through the buckle and one chinstrap attachment loop and was secured to the other loop. In all cases, this attachment was by means of an eyelet.

The liner band was made of felt or woven material. It was fixed to the helmet shell by means of four metal strips in the same way as the Adrian liner. The metal strips were spaced evenly around the helmet shell. On the early helmet, the metal strips were fixed to the helmet shell by soldering. On the later helmets, each strip was secured to the shell by two rivets of the same type as were used to secure the ventilation hole cover. The rivets in each case were 43mm apart, measured from the outside. The ends of the metal strips passed through the liner headband and were then folded flat against the helmet shell.

Between the helmet shell and the liner headband four or fewer corrugated ventilation/sizing strips were attached, as in the Adrian helmet. The metal attachment strips used to secure the liner headband also held the corrugated ventilation strips in place. On the early helmet, the corrugated strips were made from steel and not aluminium, whereas on the later helmet examined, the corrugated ventilation/sizing strips on each side were made of aluminium whilst those at the front and rear were of steel. Where four strips were present, the gap between each strip was approximately 15mm on all helmets.

On the early helmet, the liner headband was of a coarse, brown material. The liner was of one-piece light brown cotton stitched at the front and drawn together at the apex with a woven drawstring. A single row of stitching attached the liner to the liner headband. On the later helmets, the liner cloth was white or cream coloured, but other material appears to have been used. In particular, an example with blue coloured material has been noted, but this liner may well be post-war.

Other types of liner exist, although it is again impossible to say when these were fitted. They were made of leather and had a number of tongues, varying from three to eight. They were secured at the apex by a drawstring.

After the war, Sohlberg helmets were used by a number of countries apart from Finland and Russia, including the Baltic States and Poland.

Finnish Sohlberg Model 1915-17. This helmet is believed to be of early manufacture. Note the absence of rivets used to secure the liner attachment fittings on later helmets. (Collection of David Acon)

Interior view of helmet shown on the preceding pages.

Finnish Sohlberg Model 1915-17. Believed to be of later manufacture. (Collection of David Acon)

Comparison of ventilation hole covers. At top, Finnish manufacture, at bottom, Russian.

FRANCE

CHAPTER SIX

FRANCE

In common with other countries, in August 1914 the heads of the majority of French troops were covered with nothing more substantial than cloth. Some helmets were in use, for instance those issued to mounted troops, but these were not suitable for the type of the warfare that quickly developed. The helmets were conspicuous and generally unsuitable. The position was improved by the use of cloth covers.

Some helmets were modified. In particular, there was an initiative by the firm of Frank and Siraudin to modify a number of Model 1872/74 Dragoon and Cuirassier helmets. This modification included the removal of the high crest and ornamentation. In some cases, the fixing holes in the helmet shell exposed by the removal of the crest were blocked with metal disks. In other cases, the holes were covered with a leather strip or a low, Adrian-style crest. The helmets were finished by painting with a non-reflective finish. These helmets were not in use for long, the Adrian Model 1915 helmet becoming generally available.

Another attempt arising out of the urgent need for a suitable protective helmet was the re-issue of heavy helmets that had last been used by engineers during the Crimean War in 1854/55. This helmet had a pot-like shape, with a small visor and neck guard. The sides of the helmet were lower than the visor and neck guard to provide some protection to the ears and side of the head and there was a panel at the rear of the helmet constructed of thinner steel than the rest. There were twelve prominent rivets around the lower part of the helmet bowl. These secured the padded liner. In addition to the original helmets pressed into service, the steel-works at Saint Chaumond produced a new version of the old helmet. The new helmet was distinguishable from the original in that it was made from a single piece of steel, instead of in two sections, and had only five small rivets securing the liner.

These helmets weighed approximately 7kg and could only be used by troops in static positions. Indeed, the helmet was so heavy that a special supporting frame was devised. This consisted of two articulated clamps. These were attached to each side of the helmet and were adjusted so as to distribute the weight of the helmet to a frame resting on the wearer's shoulders. The helmets saw only limited service.

Another early attempt to introduce protective headwear was a commercially produced metal liner designed to be worn inside the headband of the standard cap. Few examples of this protection are known. It was not generally distributed to the troops, only being available by way of commercial sale. For further comments in respect of this protection, see under Great Britain.

However, the helmets that were in use, or were put into service, were still not suitable, particularly in circumstances where mounted troops soon found themselves fighting on foot and involved in trench warfare. In addition, there was no protective helmet available to the vast majority of the troops. As casualties rose, the need for such a helmet quickly became urgent.

The "Secrete" or "du Commerce"

The French made use of a privately-purchased head protection which was very similar to the Anglo-French Steel Cap Liner described in the section under Great Britain, the British version being almost certainly inspired by the French design. The main difference between the two designs lies in the fact that the French version does not have a rear section, but is rather designed to be held in place by the service cap, or kepi, with provision for attachment of a strap passing to the rear of the head. It has been suggested that its inventor was a chemist in the town of Chateauroux. The example examined is made of three pieces of sheet steel welded together. They comprise: the headband, the top front and top rear. Overall measurements are: 163mm front to rear, 161mm wide and 58mm at the sides. The strap attachment slots measured 18mm high by 4.5mm wide.

The Cerveliere or Calotte (Skull Cap)

The introduction of the skull cap, or Cerveliere as it was known, marked the first large-scale issue of protective headwear during the Great War. The Cerveliere was nothing more than a steel skull cap consisting of a demi-sphere made of steel .5mm or, in some cases .7mm thick, and designed to be worn under the kepi, the Chechia (cloth headwear worn by Zouaves and Tirailleurs) or other headwear.

Following successful experiments, in February 1915 an order for some 700,000 Cervelieres was placed with the Japy Company. Delivery was to commence on the 10th March 1915 at the rate of 20,000 per day. The distribution was limited to front-line troops. The Cervelieres were delivered in boxes containing 200 each. Notices accompanying each box gave details as to the method of use. The Cerveliere was to be placed in the kepi or other headwear, inside the liner band. Two small holes opposite one another enabled the Cerveliere to be fixed in place by stitching.

There were three sizes of Cerveliere. The smallest size was marked with a blue paint mark and the largest with a red mark, the middle size being unmarked. By the 27th March 1915 some 207,000 had been delivered. Unfortunately, they were all of the middle size (180mm diameter). Subsequent shipments comprised 15% of the small size (170mm diameter) and 15% of the large size (190mm diameter).

Although the wearing of the Cerveliere was compulsory, and its use undoubtedly saved lives, it was unpopular with the troops. It was difficult to fit into the kepi or other headwear and was a poor fit on the head. The lack of ventilation caused sweating. Experiments to improve the design by making the Cerveliere oval, and providing a ventilation hole in the top, did not assist. To avoid the problems, as is shown by contemporary photographs, some soldiers wore the Cerveliere on top of their kepi where it was less effective. Others wore it without any other headwear.

The Cerveliere remained in use until the Adrian Model 1915 helmet was introduced. Thereafter, production was terminated. The Cervelieres were put into use for cooking and similar purposes and to contain cartridges in the front line.

The French Adrian Model 1915 Helmet

Design, Introduction and Production

The next step was the introduction of the Adrian Model 1915 helmet. This followed a competition whereby designs were invited from various individuals. The design and introduction of the helmet is attributed to L'Intendant General A. Adrian although, without doubt, others played a part. It cannot be said that the Adrian Model 1915 helmet was a particularly innovative design. The shape, in general terms, bore considerable resemblance to earlier helmet designs, particularly the helmet used by the Fire Service Models 1885-1895-1908.

In any event, the initial Adrian Model 1915 helmets were ordered in spring 1915. The first 1,000 helmets were delivered in July 1915. By the start of the Champagne offensive on the 25th September 1915, all troops involved in the offensive were equipped with the helmet. In six months some 2,000,000 helmets were provided and, by the end of the war, approximately 20,000,000 helmets had been manufactured. French manufacturers supplied helmets not only to their own armed forces, but to the armies of many allies.

When the placing of orders for the helmets was initially considered, it was felt that the manufacture of the huge number of helmets required at an acceptable price was beyond the capabilities of the usual helmet manufacturers, namely Siraudin & Co. and Frank & Helbronner. The initial orders were therefore placed with Auguste Dupeyron and La Compagnie des Compteurs et Materials d'Usines a Gaz. Both of these companies were based in Paris. Later, helmets were manufactured by other companies. A total of eight manufacturers are known to have produced the Adrian Model 1915 helmet. They were as follows:

- Les Etablissements Dupeyron in Paris.
- La Compagnie des Compteurs et Materials d'Usines a Gaz de Paris.
- La Compagnie Coloniale.
- La Societe des Phares Autoroche.
- Le Jouet de Paris.
- Les Arsenaux de Brest et Cherbourg
- Les Etablissements Japy a Fesches-le-Chatel, Doubs.
- Les Etablissments Reflex.

There were small differences between the helmets made by the different manufacturers, for instance the crest varied in minor detail. Some helmets were marked, using an ink stamp, with the name or logo of the manufacturer. However, there appears to have been no uniform approach and many helmets were unmarked, or at least no marks are visible today.

There were some negative aspects to the helmet including the somewhat limited protection it afforded, the inconvenience created by the neck protector when a pack was being carried and the difficulty caused when wearing the helmet whilst shooting from the prone position or using a periscope. However, the speed with which the helmet was introduced and issued to the troops, its popularity with the troops, the modest cost of manufacture and the lightness of the helmet when compared with the helmets of other nations, more than compensated.

The Manufacture of the French Adrian Model 1915 Helmet

The helmet was manufactured in four main sections, namely the bowl, the crest or comb, the visor and the neck protector. The sheet steel used was without flaw or cracking and was .7mm thick. The steel was provided from two main sources, namely:

- La Societe des Forges de Franche-Comte (Usine de Fraisans-Jura et de M. M. Campionnet-Saone et Loire)
- Les Forges d'Audincourt (Doubs)

The various components of the helmet were first stamped from sheet steel. Once stamped, the metal was cold pressed to shape. The helmet bowl base, as pressed, was circular. It was necessary to make the bowl oval-shaped by forcing it over an appropriate shaped block of wood. The helmet bowl was pierced at the apex with a ventilation slot the sides of which were raised by 4mm, or thereabouts. Two rectangular holes were cut in the front of the helmet bowl to enable the badge to be fixed. The holes measured 5mm x 1mm and were 35mm apart with the lower hole being 28mm from the base of the helmet. Four holes were also made in the helmet bowl to accommodate the split pins that secured the crest.

The components were then assembled to form the helmet. The front visor and neck guard fitted into a groove at the base of the bowl and overlapped on each side. On early helmets, the overlapping sections of the front visor and neck guard were fixed only by welding, brazing or soldering and crimping the metal along the bottom edge on each side. This was found not to be entirely satisfactory and, from 1916 onwards, the overlapping sections were normally fixed by two small rivets on each side. The rivets were usually placed one above the other. However, variations existed, the most common being where the rivets were placed diagonally, the upper rivet being further to the rear of the helmet than the lower. A helmet has been examined where there is only one rivet. In addition, on some helmets the metal of the visor overlapped the neck guard on each side of the helmet whilst, on other helmets, it was the other way round. These variations were probably due to the preference of individual manufacturers.

The visor formed an angle of 22 degrees and the neck guard an angle of 45 degrees from the horizontal base of the helmet bowl. For a size A, the smallest size of helmet shell, the visor was 5cm deep and the neck guard 4.5cm deep at their maximum. The increase in size of the visor and neck guard for the larger sizes of helmet was in direct proportion to the increase in height of the helmet bowl. In order to provide a smooth edge, approximately 3mm around the visor and neck guard was folded upwards and pressed flat.

The ventilation slot cut in the apex of the bowl was covered by the crest which, in turn, had an additional slot on either side of the ventilation slot where the metal was bent outwards to allow the free passage of air. The crest or comb of the helmet was fixed by means of four aluminium split-pins using the pre-cut holes. However, it is perhaps worth mentioning that two helmets have been examined with steel split-pins.

The manufacturing tolerances (taken from *Bulletin Officiel du Ministere de la Guerre. Description des Uniformes*, 1923) for the helmets were as follows:

- 2mm plus or minus for dimensions of 170mm or more.
- 1mm plus or minus for dimensions of 100mm to 170mm.
- .5mm plus or minus for dimensions of 40mm or less.

The thickness of the sheet metal could vary by up to 5% more or less. The metal wire of the chinstrap attachment could vary by up to .5mm more or less. The angle of the visor and neck protector could vary by up to 3 degrees one way or the other.

The weight of the complete helmet varied according to the size. The size A weighed about 640 grams and the size C about 760 grams.

The Liner

Around the interior base of the helmet bowl there were up to four rectangular strips of corrugated aluminium. These measured .20mm thick and 3.5cm high. The aluminium strips served to improve ventilation and provided a means of fine adjustment of the size of the helmet. The gaps between the strips varied from 5mm to 15mm depending on the helmet size. The number of the strips and the depth of corrugation also varied according to the helmet size.

The corrugated aluminium was fixed by means of four metal strips which were soldered at the front and rear and to either side of the helmet. These also served to fix the liner. Those on either side also formed the base of the chinstrap attachment. The fixing strips were 70mm long and 5mm wide. The central section of these strips was soldered to the helmet 10mm above the groove at the base of the helmet bowl. The free ends of the metal strips were folded inward. The twin prongs thus formed passed through the aluminium strips and the cloth of the liner headband. They were then folded flush with the helmet bowl and were covered with the leather part of the liner.

The liner consisted of a band of felt or, more usually, cloth which was 4cm wide for all sizes of helmet. This went around the interior base of the bowl and was fixed to it, as mentioned above, by the four double prongs. The cloth band was made up of two layers of cloth forming a thickness of 4mm to 4.5mm. Alternatively, a single similar thickness of felt was used. The ends of the felt or cloth headband were stitched together with the stitching being placed to the rear of the helmet so as not to cause discomfort to the wearer.

The type and colour of the felt or cloth used for the liner bands was very variable. Much of the cloth was surplus material left over from uniform manufacture. For instance, the bright red material used in the manufacture of trousers for the army appears to have been used in some liners, possibly from left-over material or after the use of that particular material was discontinued for the manufacture of trousers. Examination of liner headbands also reveals material of the type used to manufacture greatcoats and the short, black uniform jacket. Also sometimes seen is finer material of the type used in the manufacture of officers' uniforms. The colour of the cloth varies from bright red to dark red, light blue to dark blue and from brown to black.

The leather liner was stitched to the liner headband. Prior to September 1916, liners were of one-piece construction with either a black, shiny finish or in a natural brown. The liner had seven tongues meeting at the apex of the helmet bowl. However, it was recognised that there was considerable wastage of leather when making a liner in one piece. A decision was reached in September 1916 to cut out the liner headband and liner tongues as separate pieces and then stitch them together. Small pieces of leather, which otherwise might have been wasted, could thus be used to make the individual liner tongues. The leather used in the later liners may be brown or black or a mixture of the two. The later liners had six tongues. On all liners, each tongue had a hole near the end, reinforced by a metal eyelet, and through which a cotton lace was threaded and tightened to adjust the fit of the helmet. The leather used for the liners was normally sheepskin.

Sometimes a liner may be encountered where the headband and tongues are separate pieces of leather but there are seven tongues. These liners are not French. They are usually of American or Belgian origin.

The Chinstrap

The chinstrap was generally made of brown or tan goatskin. It was 15mm wide, 570mm long and a minimum of 8/10mm thick. The chinstrap had a buckle without prong to provide for adjustment. The buckle measured 23mm square and was made of steel, painted black. The chinstrap was attached to the helmet on each side by means of threading it through a rectangular attachment point.

To attach the chinstrap, one end was passed around the centre strut of the buckle and secured. The other end was passed through the right side attachment loop, through the buckle and then through the left side attachment loop. A short section was then folded back against the chinstrap and secured. The chinstrap was attached permanently to the helmet. There were three methods used to secure the chinstrap to the buckle and the helmet at each end. They were:

- The first method used, and the most common, was by means of a copper eyelet and washer. The eyelet had a hole in the centre open at both ends.
- A brass eyelet, smaller than the first type used, with a washer and hole on one side but closed on the other side.
- A small rivet in the form of a split-pin. This method is thought to be the last method used.

Other types of chinstrap may be encountered. Often these are post-Great War additions. However, simply because a chinstrap is not of the standard type it should not be assumed that it is not of the Great War period. Chinstraps were undoubtedly replaced in the field, and the replacement was not necessarily of regulation type. Additionally, and more usually, chinstraps were purchased privately, particularly by officers. They were of various designs and were usually, but not always, of plaited leather. They were generally of two-piece construction secured by a buckle with a prong. They were often attached to the helmet by means of a separate leather loop and stud arrangement on each side of the helmet.

The Chinstrap Attachments

The chinstrap was attached to the helmet by means of a rectangular wire loop on each side of the helmet. The rectangular attachment loop was 17mm long and 7mm wide and was made of 2mm diameter wire. The attachment of the rectangular loop was by means of an extension of the metal strip used to fix the ventilation strips and liner. It was 0.5mm thick, 11.5mm wide and 15mm long. The lower part of this fitting was wrapped around the rectangular attachment loop and the upper part was soldered to the interior of the helmet bowl.

Other styles of chinstrap attachments exist. The most common of these was where the chinstrap attachment loop was attached directly to the underside of the helmet brim. Attachment was by means of the two rivets used to secure the join between the visor and the neck guard. It is said by some that this form of chinstrap attachment was used on early helmets. Others have commented that this form of chinstrap attachment was a post-war alteration. The authors comment that they have examined six Adrian Model 1915 helmets where the chinstrap attachment is by means of two rivets to the underside of the joint between the visor and the neck guard. The helmets were four with French badges, one Czech Legion and one Belgian. None of the helmets have the features common to early production helmets.

The chinstrap attachments on the helmets examined were all identical, the rectangular wire attachment loop being 21mm wide. The small piece of metal securing the rectangular attachment loop to the helmet in all cases had a rounded end and was 13mm wide. Only one end was visible, the other end being sandwiched between the helmet bowl and visor. The helmets were all very similar indeed, the crests being of identical shape. On all the helmets there was a feature not usually seen in that the visor and neck guard were spot welded to the helmet bowl using three welds to the visor and three to the neck guard. The liner fixing points were all spot welded in three places to the helmet bowl.

All the helmets examined appeared to have an original Great War paint finish. The Belgian helmet had some small initials and 1918 scratched on the underside of the neck guard. All the helmets showed considerable wear, save one. That exception had a matt, mustard paint finish. Two of the helmets had a manufacturer's ink stamp consisting of an anchor and two Cs. This is thought to be the stamp of la Compagnie Coloniale. In short, all the helmets appeared to have been produced by the same manufacturer. It is almost inconceivable that all the helmets were altered in an identical way after the war ended. The only conclusion is that they were a variation of the standard method of chinstrap attachment. The modification may have been necessary to strengthen the chinstrap attachment. In that respect, a number of Adrian Model 1915 helmets have been examined where the chinstrap attachment had broken and was soldered back in place.

Paint Finishes

The history and study of the various paint finishes found on Adrian Model 1915 helmets is, frankly, somewhat confusing. During the Great War, a number of paint finishes were used and the shade and finish of the paint varied enormously. Different manufacturers used different shades and finishes of paint. Following the introduction of the various official changes in the shade or colour to be used, many helmets were repainted at unit level with whatever suitable paint was available. Over the years since the paint finishes have aged in different ways, getting darker or lighter as the case may be, depending on the type of paint used originally and the circumstances of storage since.

To complicate matters, following the Great War the French Army as a whole adopted khaki as a standard uniform colour. Although new models of helmets were adopted, starting with the Adrian Model 1923, some Adrian Model 1915 helmets remained in use until and during the Second World War. Some were repainted to match the new uniform colour. Thus, it is not uncommon for one helmet to have a number of coats of paint. In some cases, all these layers of paint were applied during the Great War. In other cases, the final layer(s) was applied after the end of the war. It is therefore necessary to bear in mind that it is extremely difficult in some cases to be certain as to the period when a given paint finish was applied.

Generally speaking the manufacturers' system of painting used compressed air. This produced a smooth, even finish without runs or drips in the paint. Following painting, the helmets were baked in an oven at between 120 and 140 degrees Farenheit for one and a half hours. Factory painted helmets with their even finish are easily distinguishable from helmets that have been repainted by hand at unit level where the brush strokes are usually visible.

Initially, the helmets were painted with a shiny, bright blue paint. This finish was certainly attractive but was far from ideal, being both bright and reflective and thus easily visible to the enemy. To remedy this problem, some troops smeared mud onto their helmets as a camouflage finish. This method of camouflage is particularly noticeable in original photographs where chalk predominated in the soil of the battlefield. The practice was banned as the mud on the helmet surface might be carried into a wound by a projectile penetrating the helmet, resulting in infection. Instead, the use of helmet covers, mentioned below, was authorised.

In view of the urgency to issue helmets to the troops taking part in the Champagne Offensive in September 1915, they all received blue-painted helmets. Thus, the Zouaves and Tirailleurs received blue-painted helmets with the standard, flaming grenade badge. In late September 1915 helmets with the new badge for those troops were issued, but still painted blue. Subsequently, pursuant to an order dated the 31st October 1915, their helmets were re-painted brown. The shade of the brown paint used varied from a light biscuit colour through greenish-brown to a rich earth-brown.

On the 3rd June 1916, the use of a darker blue was introduced. Helmets already issued were repainted by the troops themselves. It is thus not uncommon to see the early, bright blue helmets with a coat of darker blue paint on top. Factory painted helmets continued to vary significantly in the shade of paint used and the texture of the finish.

On the 25th May 1918 a decision was taken that colonial troops would also adopt the mustard colour for their helmets and helmets produced after that date were painted in that colour. The existing blue helmets were repainted. When helmets were repainted, almost invariably the interior of the helmet bowl was left in the original colour.

Thus, to summarise, helmets may be painted with a shade of blue that varies from a light, shiny blue to a midnight blue bordering on black. There may be more that one coat of blue paint with, usually, the darker shade painted on top of the lighter shade. Helmets that were originally blue may have a brown coat(s) on top of the blue. The brown paint may vary greatly, as mentioned above. Khaki paint applied after the Great War is sometimes difficult to distinguish from the earlier finishes. The post-war paint is likely to have a matt finish and to be of a khaki colour rather than mustard or of a more green or brown shade.

Helmet Covers

An order of the 30th October 1915 authorised the use of cloth covers. Initially, the order anticipated that the covers would be in an appropriate blue colour. Needless to say, various types of cloth were used and the colours included blue, brown, beige, white, cream and grey. The use of cloth covers appears, from contemporary photographs, to have been quite extensive. Cloth covers continued in use even after helmets were finished in darker, matt paint finishes.

The covers were normally secured on the underside of the helmet brim using a cotton drawstring. However, covers fitted with metal clips to secure the cover to the helmet are known.

Helmet Sizes

The actual helmet shells were produced in three basic sizes, designated sizes A, B and C. The helmets themselves were restricted to three basic sizes in order to simplify manufacture by, for instance, keeping the required tool-making to a minimum and thus keeping costs to a reasonable level.

In addition, each of the three basic sizes of helmet was, in turn, divided into three sub-sizes. This was achieved by using deeper or shallower corrugations in the aluminium strips or by removing some of them. There were therefore, in all, nine separate sizes ranging from 54cm to 62cm.

The practice of marking helmets with their size appears to have been very variable. Some appear not to have been marked at all. In respect of others, the interior of the helmet bowl was stamped with the size, using an ink stamp. These stamps take different forms. The stamp may give the letter of the main size followed by the number of the sub-size. For instance, B2 indicates the middle size of the three size B helmets. Alternatively, they were stamped with the main size followed by the size of the liner in centimetres. For instance, C61 indicated the middle of the three size C helmets.

The size stamp was sometimes placed on either side of the ventilation slot but stamps may be found elsewhere inside the helmet bowl. As an alternative, the leather of the liner itself was stamped at the front with the size. This practice appears to apply only to the early, one-piece liners.

For each 1,000 helmets, what was regarded as an appropriate number of each of the nine sizes were produced.

The following table of dimensions of the various sizes, adapted from the publication "Description des Uniformes", may be of interest:

Helmet Sizes

	A1	**A2**	**A3**	**B1**	**B2**	**B3**	**C1**	**C2**	**C3**
Circumference of the interior of the bowl in cm.	60	60	60	63	63	63	66	66	66
Depth of helmet bowl in mm.	105	105	105	110	110	110	115	115	115
Diameter of helmet bowl front to rear in mm.	201	201	201	210	210	210	219	219	219
Diameter of helmet bowl side to side in mm.	181	181	181	190	190	190	199	199	199
Interior circumference of helmet with liner in place in cm.	54	55	56	57	58	59	60	61	62
Number of helmets per size per 1,000 manufactured	10	70	250	250	160	100	100	40	20

Helmet Badges

Each helmet was issued with a badge. The badges were stamped from sheet steel or tin plated iron of 4/10mm thickness. The badge was fixed by means of a strip of steel 5mm wide and 75mm long. Of this strip, 35mm was soldered to the rear of the badge. The ends were bent at right angles to form two prongs each 20mm long. One of these prongs passed through each of the two small rectangular holes cut in the front of the helmet. The ends were then folded inwards, flush with the interior of the helmet bowl, thus securing the badge. The badges were painted on both sides in the same colour as the helmet.

There was considerable debate as to the choice and design of the badges. For instance, it was argued that the badge for the Zouaves should feature a capital Z and that of the Tirailleurs a capital T. On the 19th August 1915, in order to resolve that issue, General Joffre ordered that they should share the same badge. The badges adopted were not, in the main, new designs. They had mostly been used, in slightly different form, on the helmets and other headwear in use in the previous century.

There were initially seven badges, with one further badge being added later during the War. The majority of the badges incorporated the capital letters RF which stood for Republique Francaise. The standard badges, including the date of introduction, were as follows:

1. The flaming grenade for the Infantry was introduced in February 1915. The use was extended to the Cavalry on the 17th June 1915, Cyclists on the 28th June 1915, Light Cavalry and Chasseurs d'Afrique on the 16th October 1915, the Gendarmerie on the 6th November 1915, Train Companies on the 2nd March 1916 and others such as the Foreign Legion and other troops without a specific badge on the 10th October 1916. The dates are the official introduction dates. Prior to that, the troops in question, if helmets were issued to them, used the badge in any event. The badge measured 63mm high and 42mm wide.

2. The hunting horn for Chasseurs a Pied and African Light Infantry. Introduced in February 1915. The badge measured 40mm high and 65mm wide.

3. The flaming grenade with crossed cannons below for the Artillery including Colonial Artillery. Introduced February 1915. The badge measured 63mm high and 80mm wide.

4. The helmet and breastplate for the Engineers. Introduced February 1915. The badge measured 73mm high and 44mm wide.

5. The crescent for Zouaves and Tirailleurs. Introduced August 1915. The badge measured 53mm wide and 44mm high.

6. The Caduceus (winged staff with entwined serpent) for medical troops. Introduced August 1915. The badge measured 70mm high and 78mm wide.

7. The flaming grenade with anchor for Colonial Infantry and the Navy. Introduced June 1915. The badge measured 73mm high and 48mm wide.

8. The badge consisting of a Laurel wreath, flags and fasces for officials in the L'Intendance (Quartermaster Corps). Introduced March 1916. The badge measured 63mm high and 65mm wide.

Small differences in the measurements of individual badges will be encountered. In addition, there were considerable differences in the manufacturing detail of the same type of badge. These were largely caused by differences in the engraving of the original dies used to stamp the badges, or by wear on those dies once they had been used for a time. This caused some badges to be very detailed whilst others were less so. However, there were other differences. For instance, the metal between the ends of the crossed cannons on artillery badges was left in place, to a greater or lesser degree, whilst on others it was absent.

Other badges will, from time to time, be encountered. Simply because a badge is not one of the standard type, it should not be assumed that it is not of the Great War period or that it is a modern copy intended to deceive. Some examples of non-standard badges which have been observed are as follows:

1. A handmade Zouaves crescent badge with the letter Z intertwined with the numeral 1 used by the 1st Zouaves Regiment.

2. A silver cross soldered onto a standard infantry badge and used by chaplains.

3. The standard infantry badge with the numeral 1 superimposed and used by the 1st Foreign Legion Regiment.

4. A badge made of brass and forming the word GAZ. This was possibly worn by troops involved with the detection of gas on the battlefield. However, there is no documentary evidence in this respect and it is perhaps more likely that the badge was worn by those involved in aspects of civil defence where gas was involved such as the guarding and protection of gas works.

5. A five-pointed star 4cm high used by cavalry reconnaissance troops.

6. The numerals 75 framed by two wings with crossed cannons above. This was used by certain anti-aircraft batteries.

7. A disc above the standard grenade badge with the letters "AS" meaning Artillerie Speciale.

8. A crescent made of copper and bearing the number 1 used by the 1st Tirailleurs Algerian.

9. Badges taken off earlier models of helmet. This has been particularly noted in respect of the artillery crossed cannon and the hunting horn badges taken from the Model 1872 Shako. It was in the main officers who wore the earlier badges, presumably to distinguish their helmets from those worn by the rank and file.

10. Gendarmes in forward areas were equipped with the standard Adrian Model 1915 helmet with the infantry badge painted white.

11. It should perhaps be mentioned that the badge consisting of an anchor only, without the flaming grenade, was not adopted for the Navy until 1929. However, that does not mean that a badge consisting of an anchor only and made of copper or brass was not fixed to a helmet during the Great War. Such a badge may have been removed from a Model 1896 colonial helmet.

12. Painted badges consisting of one of the badges of a suit from a pack of cards have been observed. These were probably worn by members of armoured units.

13. A standard flaming grenade badge, with a pair of wings made of brass attached. This badge is thought to have been worn by a member of an aviation unit.

14. The red cross emblem. This might be worn by stretcher bearers, nurses and other medical personnel. It is, of course, very difficult to say when such a badge was applied to the helmet as the badge continued in use after the war ended.

The above list is far from complete and other individual and often hand-made badges will be encountered. For instance, in the photographic section there is a example of a copper badge in the form of a trench mortar bomb.

In addition, rank insignia were sometimes fixed or painted to the front or side of the helmet. The insignia may take the form of metal bar(s) or white metal star(s). Some officers wore pre-war badges in the form of the flaming grenade but lacking the letters "RF". The troops of the Chasseurs a Pied sometimes wore their helmets painted very dark blue or black with the hunting horn of the badge left as bare metal or highlighted in some other way.

Helmets Issued to Non-Military Personnel

Adrian Model 1915 helmets were issued to non-military personnel serving near the front in areas liable to artillery bombardment. These personnel included mayors and civil servants and similar who received a helmet without a badge or badge fixing holes.

Fire-fighters near the front received standard infantry helmets with the badge painted red and labourers under military command wore a helmet with a badge consisting of a disc 35mm in diameter with the letters RF punched in it.

The French Adrian Model 1915 "Artillery Special"

The first tank units were created at the end of 1916. They were initially known as "d'Artillerie Speciale" or "d'Artillerie d'Assaut".

The title "d'Artillerie Speciale" was generally adopted, shortened to "AS" and that title was adopted because most of the tank troops involved came from the artillery.

With the introduction of the tank, and particularly of the small two-man Renault tanks where space was critical, it was realised that a special helmet was needed for tank crews. The brim of the standard helmet interfered with the ability to look through observation slits and periscopes.

The simple expedient was to cut-down the brim. This conversion was carried out at unit level. The method of conversion most commonly adopted consisted of removing the brim from just in front of the join between the visor and neck protector. The brim of the helmet left was only 1.5cm deep.

In some cases, the conversion included the addition of cloth or leather covered padding to the front of the helmet where the visor had been cut down. The padding was in the form of a roll, rather than being made up from separate layers. Improved protection was occasionally provided by the insertion of layers of cloth padding into the liner band.

Another method of conversion occasionally used was more radical. It involved the removal of the visor and neck protector and crest completely, leaving only the helmet bowl. The top of the helmet was covered with a strip of leather, in effect replacing the crest.

Troops entering the armoured units retained their original helmets, usually with the artillery badge. However, troops from other units were involved and they retained their distinctive helmets which were converted in the same way. The converted helmets are known as the "Artillery Special".

The converted helmets were worn together with the British leather-covered steel and chain- mail face protection, stocks of which were provided by the British. For more detail in this respect see under Great Britain.

In the second half of 1918, Colonel (later General) Estienne, who is widely regarded as the "father" of the French armoured units, gave a direction for a new helmet to be produced. This helmet was designated the Model 1919. It was a helmet produced specifically for the crews of armoured vehicles, and was not a conversion.

It is just possible that, despite the designation Model 1919, the helmet saw service at the end of the Great War. Additionally, it is appropriate to highlight the differences between the two helmets. Features found on the Model 1919 are as follows:

- The crest was shortened.
- The neck guard sloped downwards at a steeper angle.
- There was a flat, rather than rolled, front pad. This was made up from three layers of leather, fixed by rivets on each side. The bottom of each layer of leather padding was folded inwards and upwards.
- The badge fixing slots, designed to take the new larger tank insignia, were higher on the helmet.
- The aluminium ventilation spacers covered the entire interior circumference of the helmet bowl, rather than leaving gaps between them.

Other Helmets and Developments

Although of attractive design and popular with the troops, the Adrian Model 1915 helmet was of light construction and, in reality, provided only modest protection to the head and no protection to the face and eyes. In common with other countries, the French made considerable efforts to provide their troops with greater protection. Some of that protection was provided in the form of body armour. Additionally, loophole plates which could be either fixed in a trench or worn as a form of armour were quickly developed. These go beyond the scope of this book.

As far as helmets were concerned, there were numerous experiments with helmets being produced and used in the front line. However, ultimately most of these experiments were unsuccessful. This was mainly due to the unacceptable weight of a helmet offering greater protection. Alternatively, there was interference with vision in respect of those designs offering protection to the face and eyes.

The following are examples of the attempts to offer greater protection.

The Brow-plate

These were similar in concept to the German "Stirnpanzer", being fitted to the front of the helmet and secured by a leather strap around the rear of the helmet bowl. The metal was 4mm thick and the weight 2.5kg. The brow plate appears not to have been used to any great extent. Certainly original French Adrian brow plates are far more scarce today than their German counterparts.

The Polack Visor

Following acceptance of the Adrian Model 1915 helmet, efforts were directed towards developing a visor to fit to the standard helmet to provide protection to the face and eyes. The French Bureau of Inventions, a sub-section of the Department of War, considered various designs. Major Polack worked in the Bureau and he designed a number of visors. In most of the Polack designs, the helmet could be worn with the visor raised on the front of the helmet or lowered to partially cover the face and eyes.

The earliest design attached to the helmet by means of a rubber band. Another, probably late, design was made to attach neatly to the helmet visor. Other designs provided for attachment of the visor to the helmet on either side, close to the chinstrap attachments.

Initially, vision slots were cut through the metal of the visor. In later versions, vision was provided by a series of metal strips with gaps between and built up in a framework which was then inserted in the body of the visor. The metal strips in front of the eyes were horizontal whilst those above slope upwards and those below slope downwards. The metal strips were of thin metal and, being placed edgeways to the line of sight, allowed reasonable vision and protection. However, it was found that the angling of the strips, designed to improve vision, tended to channel low velocity fragments towards the eyes. In addition, as a defence against a projectile travelling at moderate velocity, such as a pistol bullet, the visor was worse than useless. The reason for this was that the projectile was likely to be deformed as it passed through the visor leading to a more severe injury. Additionally, parts of the visor might bend or break inwards, again resulting in a more serious injury.

Production of helmets fitted with Polack visors was modest, most being manufactured for experimental or trial purposes only. However, in early 1918, several thousand of each of two models were manufactured.

Other Helmets used by the French

The Dunand Helmets

John Dunand was, in fact, a Swiss national. He volunteered for the Red Cross and served on the Western Front as an ambulance driver.

The couturier Jean-Philippe Worth had converted his salon at 7, Rue de la Paix, Paris into a hospital. Worth had an interest in the treatment of head wounds, and the hospital specialised in that treatment. Dunand had an interest in protecting the soldiers from such wounds, and particularly from injury to the eyes. It was thus arranged that Dunand would go to work at the hospital.

Dunand produced privately a series of helmet designs. He was assisted by his brother, who had suffered a head injury in the early part of the war. Many of the designs had a visor attached. The helmets were hand-made from manganese steel. About twelve designs were submitted to the French and American authorities for consideration.

In August 1917, the American Government ordered 10,000 of the helmets to be manufactured by Crosby and Company of Buffalo, New York. Unfortunately, the original helmets had been hand-made and it was found that the design could not be mass produced in the metal specified. Ultimately, the contract was cancelled. None of the Dunand designs was accepted for general service by any combatant country during the Great War.

Privately Purchased Helmets

Prior to and following the introduction of the Adrian Model 1915 helmet, private manufacturers of military equipment produced for sale, a number of helmets which, whilst having the general appearance of the standard helmet, were quite different in detail or materials used. Some of them are as follows.

The Acier Trempe

This was a helmet produced by the firms of Siraudin & Co. and Franck & Co. The shape of the Acier Trempe, or tempered steel helmet as it was generally known, closely resembled the standard Adrian Model 1915 helmet when viewed from a distance. However, the Acier Trempe differed significantly in a number of ways, as follows:

- The visor and rear brim were less angled and are pointed at the front and rear.
- The crest was more pronounced, the front being truncated and rounded.
- The interior ventilation was provided by two round holes instead of a slot.
- The holes at the front for fixing the badge were round instead of rectangular.
- The interior of the helmet was painted black. The black leather liner had a gold border.
- The chinstrap usually had an officer style buckle.
- The manufacturer's mark, consisting of a helmet with a mane-like crest, and its initials were stamped on the top centre of the rear brim.

The Acier Trempe helmet was available commercially shortly before the introduction of the new Adrian Model 1915 helmet, having been hastily marketed by the manufacturers. The helmets were purchased privately and put into service, particularly by officers.

However, the use of the helmets was prohibited, pursuant to an order dated the 29th September 1915. The prohibition was ordered because the helmets were made from tempered steel. This shattered when struck by a projectile, rather than distorting and absorbing at least some of the shock. The fragments of the shattered helmet shell tended to cause additional injury to the wearer. The prohibition may also have been imposed because it was considered the use of the Adrian Model 1915 helmet by the troops would not be encouraged by allowing officers to use privately purchased helmets. Following the prohibition, all ranks used the same helmet.

Once use of the helmets was prohibited, judging from the photographic evidence, many of the Acier Trempe helmets continued in use in photographers' studios.

Aluminium Helmets

These were almost identical to the standard helmet, save the helmet bowls were more rounded. They lacked the two rectangular slits to attach the badge and had two round ventilation holes in the apex of the helmet instead of a slot. The helmets were sometimes painted blue or olive green, but other paint finishes and helmets with badges attached may be encountered. Sometimes they are found painted for the French Red Cross. These helmets may have been manufactured for the Italians or the Americans.

Non-Metallic Helmets

These were generally made of cork with a cloth covering which was glued on and painted with a thick layer of khaki varnish.

However, an example painted blue has been observed and is shown in the photographic section. The crest was made of aluminium. These lightweight helmets are likely to have been used for parade purposes.

There was a second type which had stitching visible on the exterior and a matte paint finish. These helmets were probably issued to troops in the tropics. As with all lightweight helmets the question as to their date/period of production remains to be answered.

The Souvenir Plaque

By a decree dated the 18th December 1918, the grateful French Government awarded to each officer and soldier of the armed forces a helmet as a souvenir. The initial decree provided that each helmet would have an engraved inscription on the top surface of the front brim, giving the name and rank of the soldier with the words "Soldat de la Grande Guerre" underneath. A further decree provided that a helmet with a similar wording engraved on it would be sent, on receipt of a request, to the family of all those who had died whilst in military service.

The practical impossibility of engraving all the helmets was quickly recognised and the souvenir plaque was designed to replace the engraving. The plaque was made of embossed brass and was designed to be attached to the visor. The centre of the plaque had an oval shaped shield, the top half of which was left blank so that the recipient might have his name and other details inserted if he wished. The bottom half of the shield was embossed with the wording mentioned above. On either side of the plaque appeared Laurel leaves. The plaque was attached to the helmet by fixing at each side of the visor. Usually holes were made through the visor for this purpose, although one helmet has been encountered where the attachment is to the chinstrap using two small buttons. The attachment to the visor may be by rivets, small nut and bolt or split-pin.

Unfortunately, what appears a good idea, and a gesture which would have been appreciated by the troops, was not put into effect promptly or uniformly.

Post War Use

At the end of the war, there was a shortage of helmets for use by the Fire Brigade and other municipal services. A number of Adrian Model 1915 helmets were converted for use by these organisations. Various paint finishes and badges, either painted or in metal, were applied.

Further, the Adrian Model 1915 helmets saw considerable use in the armed forces of various countries after the war. They continued to be used by the French in considerable numbers until the end of the Second World War.

It is perhaps worth mentioning that none of the Adrian Model 1915 helmets in French service during the Great War had any rivets (as opposed to split-pins) passing through the helmet shell. With the exception of the wartime conversion of Adrian Model 1915 helmets supplied to the Italians and captured by the Austro-Hungarians, where a helmet is found with rivets passing through the helmet shell, this is likely to be a post-war modification.

The Secrete, or Du Commerce. (Collection of Brana Radovic)

Cerveliere or Calotte. The plaque has been added and bears the inscription, "La Guerre - 1915- (French Trench Helmet)". This example has been ovalised and measures 175mm from side-to-side and 190mm from front-to-back. (Collection of the Imperial War Museum, London)

Cerveliere in place inside the kepi. Note battle damage. (Collection of Brana Radovic)

16

16

Acier Trempe (Tempered Steel) Helmet. Note the early brown finish and pre-war style infantry badge without initials "RF". (Collection of Marcus Cotton)

Interior of helmet shown on previous pages.

Acier Trempe Helmet in horizon blue with pre-war style infantry badge without initials "RF". (Collection of Ellwood von Seibold)

Acier Trempe Helmet in horizon blue with officer's chinstrap and pre-war style Chasseurs' badge. (Collection Polish War Museum, Warsaw)

■ Note Acier Trempe manufacturer's mark on neck guard.

Acier Trempe Helmet in horizon blue with pre-war style Engineer's badge. Again note absence of initials "RF". (Collection of Marcus Cotton)

Acier Trempe Helmet in a brown finish with brass pre-war style badge without the initials "RF". (Collection of Ellwood von Seibold)

A standard Adrian Model 1915 Helmet. Painted in horizon blue and carrying the correct wartime steel infantry badge. Note on this example the neck guard overlaps the visor and they are secured in place by two rivets, one directly above the other. (Collection of Ellwood von Seibold)

Component parts of an Adrian Model 1915 Helmet. (Collection of Imperial War Museum, London)

Note the chinstrap attachment is fixed directly onto the join between the neck guard and visor rather than onto the helmet bowl.

Top: Standard chinstrap attachment to helmet bowl. (Collection of Michael Haselgrove)

Bottom: Note the chinstrap attachment directly to the join between visor and neck guard. Note also the spot welding securing the visor and neck guard to the helmet bowl. (Collection of Michael Haselgrove)

Top: A brown-painted Adrian Model 1915 Helmet with original blue interior. Note the rear fixing for the liner headband, the manufacturer's ink stamp, believed to be La Compagnie Coloniale and the helmet size (B2) stamped in ink on either side of the ventilation slot. (Collection of Michael Haselgrove)

Bottom: Note the fixing of the aluminium ventilation/sizing strip and one prong of the badge fixing. (Collection of Michael Haselgrove)

First pattern one-piece leather liners.

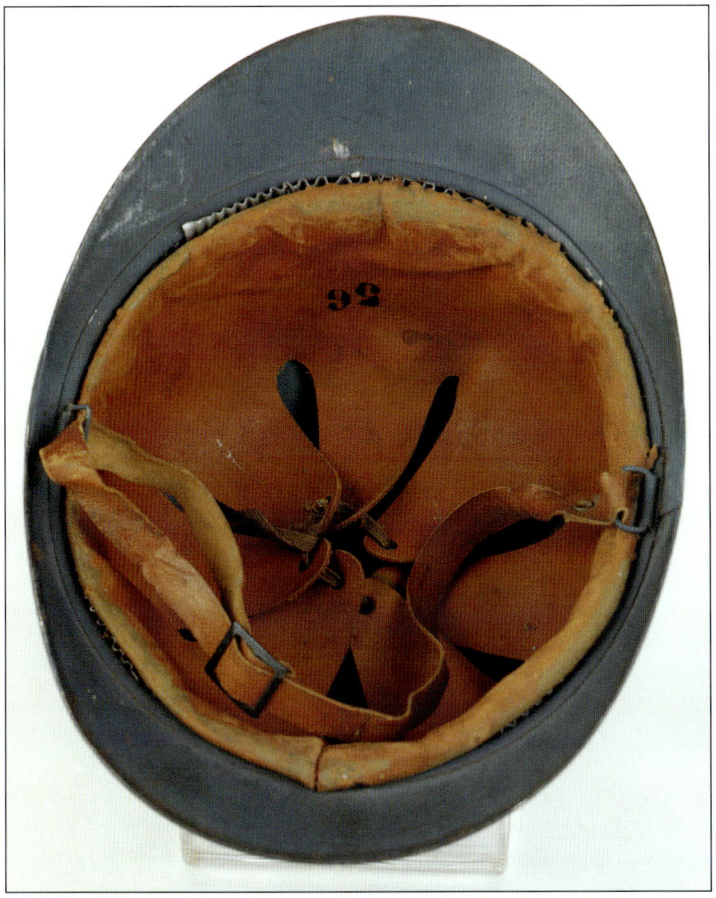

■ Top left: Brown early-style liner. Top right: Later pattern liner using brown leather only.

■ Bottom left and right: Later pattern liners using a combination of brown and black leather.

Another colourful example of the late pattern liner.

■ Top and middle: Early pattern liners showing some of the different materials used for the liner headband.

■ Bottom: Late pattern liner.

Standard Adrian Model 1915 Helmet badges. Top left: Infantry. Top right: Chasseurs. Bottom left: Artillery. Bottom right: Engineers.

Top left: Zouaves and Tiraillers. Top right: Medical. Bottom left: Colonial Infantry and Navy. Bottom right: L'Intendance (Quartermaster)

Differences in badge manufacture.

Badge variations.
Top left: Colonial Infantry or Naval, painted white. Probably for use by Military Police.
Top right: Chasseur, highlighted in green.
Middle left: Naval badge taken from Model 1896 Colonial helmet.
Middle right: Chasseur, highlighted by removal of paint.
Bottom left: Hand-made copper trench mortar
Bottom right: Chasseur, highlighted in black.

Standard Adrian Model 1915 Artillery Helmet with early horizon blue gloss finish. Note absence of rivets securing visor and neck guard with the neck guard overlapping the visor. (Collection of Ellwood von Seibold)

Standard Adrian Model 1915 Medical Helmet with early horizon blue matte finish. Note two rivets, one above the other, securing the visor and neck guard with the visor overlapping the neck guard. (Collection of Marcus Cotton)

Standard Adrian Model 1915 Infantry Helmet with late dark blue spray finish. Note the single rivet securing the visor and neck guard. (Collection of Michael Haselgrove)

Top left and right: standard Adrian Model 1915 Artillery Helmet with late dark blue brushed finish. Note steel pins securing the crest and absence of rivets securing the visor and neck guard. (Collection of Brana Radovic)

Bottom left and right: standard Adrian Model 1915 Infantry Helmet with early finish over-painted in dark blue. Note that the pins securing the crest are steel and are not the usual dome shape. (Collection of Michael Haselgrove)

High-quality brown gloss finished helmet with infantry badge. Possibly for use by the Foreign Legion or Colonial troops. (Collection of Michael Haselgrove)

Matte mustard-painted helmet with badge for Zouaves and Tiraillers. The inscription on the silver plaque reads: "Mr DUBEGH. CAPITAINE. SOLDAT. GRANDE GUERRE 1914-1918". The plaque appears to have been painted red at sometime indicating use by a Zouave. (Collection of Ellwood von Seibold)

Helmet, originally horizon blue, over-painted with matte mustard and carrying the infantry badge. This helmet has the veterans' brass plaque which states it was owned by "H. IENNY LEGION ETRANGERE". Note the chinstrap is absent, but the leather officers' chinstrap attachment points remain. (Collection of Michael Haselgrove)

Over-painted matte mustard Colonial artillery helmet. The veterans' plaque is inscribed "Vinchon Lieutenant". (Collection of Ellwood von Seibold)

Top: Over-painted matte mustard. It has three brass or copper bars soldered over the infantry grenade badge indicating the rank of captain. (Collection of Brana Radovic)

Bottom: Dark blue Engineers' helmet. The standard other ranks chinstrap has been plaited to resemble the officers' style. The veterans' plaque has been removed, leaving a remnant on each side under the rivets. Three painted white captains' rank bars are found on the upper right hand side of the bowl. The wearer's name and other details are scratched onto the top of the visor and neck guard. It may have had its veteran's plaque removed for use during the Second World War by an officer who subsequently inscribed it. (Collection of Michael Haselgrove)

Top: Medical General de Brigade helmet. The badge is a copper or brass pre-war type, probably off a shako. (Collection of Ellwood von Seibold) Bottom: Infantry General de Brigade helmet. (Collection of Brana Radovic)

Infantry officer's helmet with removable plaited chinstrap. (Collection of Brana Radovic)

Various styles of officers' chinstraps.

Artillery helmet with an unusual camouflaged paint finish after the German style. Note battle-damage. (Collection of Brana Radovic)

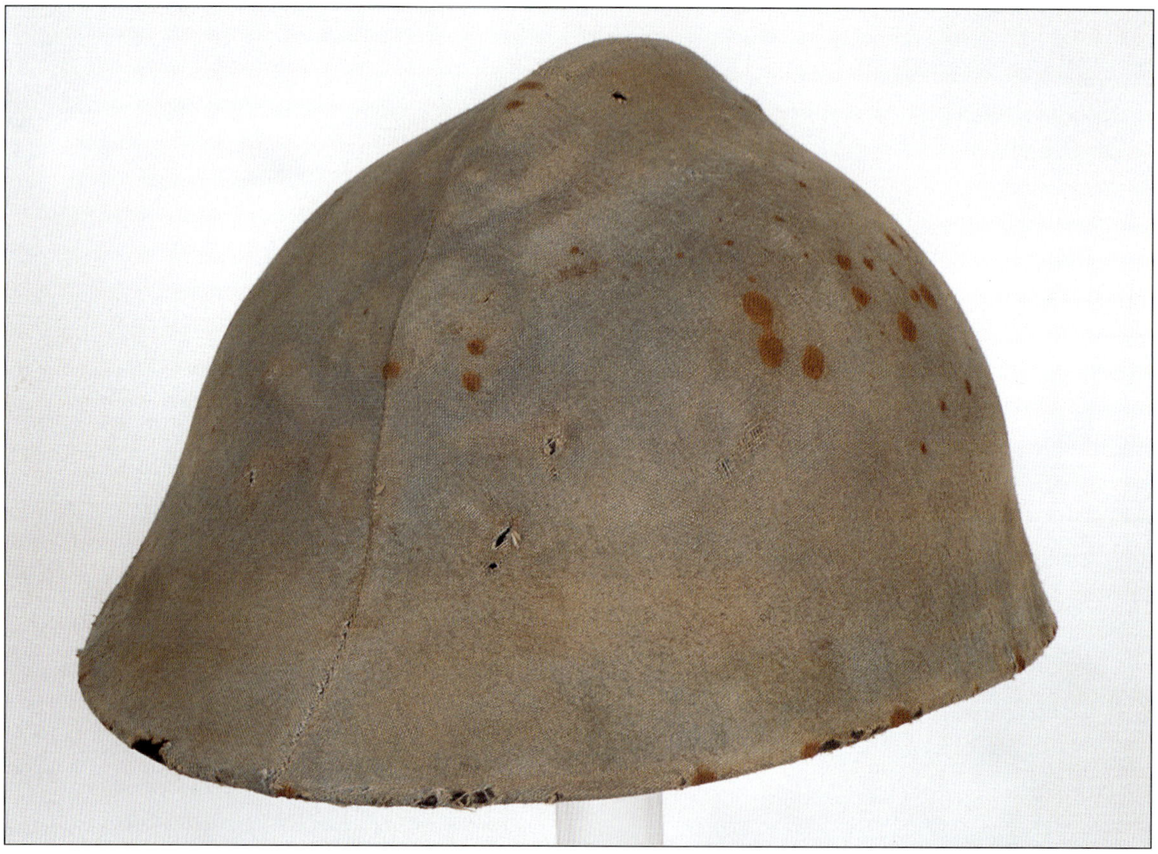

Top: Blue cloth two-piece drawstring cover with medical personnel patch applied. (Collection of Brana Radovic)

Below: The same type of cover with faded colour. (Collection of Brana Radovic)

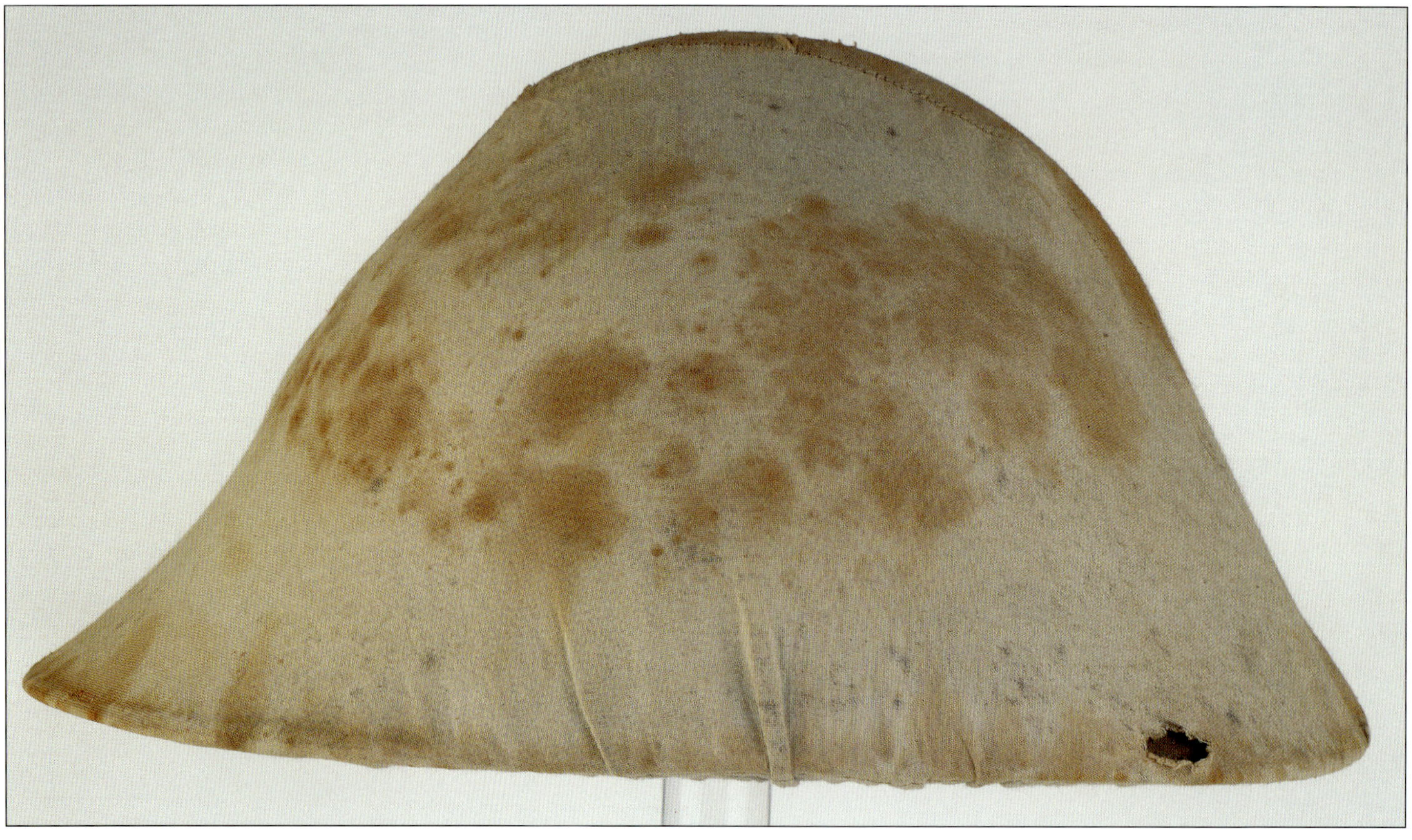

Tan cloth two-piece drawstring cover. (Collection of Marcus Cotton)

Tan heavy gauge two-piece drawstring cover. (Collection of Brana Radovic)

Damaged early Infantry helmet with three coats of paint: horizon blue, dark blue then brushed a dark green/brown. The bullet has entered front left side skirted the interior of the bowl and exited bottom rear right. (Collection of Michael Haselgrove)

Infantry helmet with shell splinter damage. (Collection of Michael Haselgrove)

Infantry helmet with shell splinter damage. (Collection of Michael Haselgrove)

Infantry helmet with bullet damage.

Colonial Infantry helmet with bullet damage. (Collection of Marcus Cotton)

Badly damaged, relic condition Colonial Infantry helmet. (Collection of Michael Haselgrove)

■ Top: Relic condition brow plate designed to fit the Adrian Model 1915 Helmet. (Collection of Ellwood von Seibold)

■ Below: Brow plate for the Adrian Model 1915 Helmet. (Collection of the Musée Militaire du Perigord, Perigueux, France)

■ Brow plate for the Adrian Model 1915 Helmet. The securing strap appears to be a post-war replacement. (Collection of the Musée Militaire du Perigord, Perigueux, France)

Adrian Model 1915 Artillery Special Helmet for tank crews. (Collection of Brana Radovic)

■ Top: Artillery Special shown with the wire mesh Sabatier Mask. (Collection of Brana Radovic)

Adrian Medical helmet with hinged Polack visor. The slots in each side allow the visor to slide and to be held in the "down" or "up" position (Collection of Brana Radovic)

R F

R F

Adrian Polack visored helmet "Long-nosed" variant. (Ellwood von Seibold)

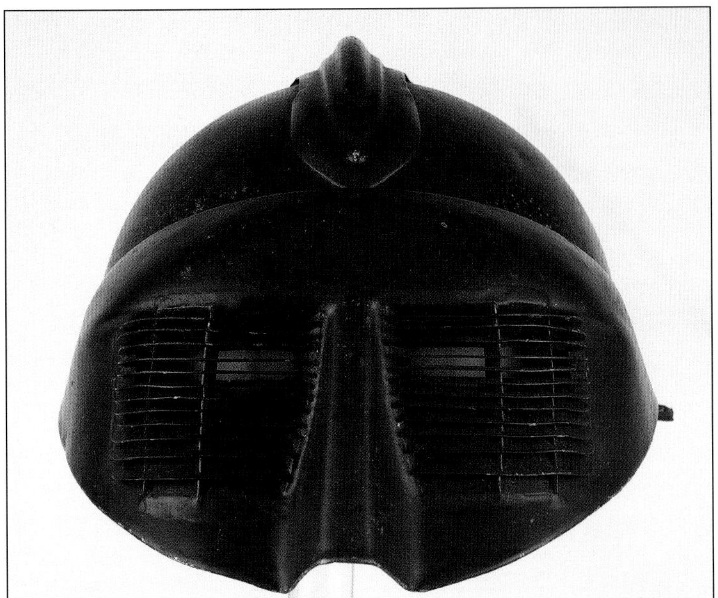

Adrian Polack visored helmet "Short-nosed" variant. (Imperial War Museum, London)

Adrian Medical Polack helmet for comparison.

Adrian Polack visored helmet. This model was designed to clip on to an issue Adrian Model 1915 helmet, thus converting ordinary helmets without the need to produce replacements. (Collection of the Musée Militaire du Perigord, Perigueux, France)

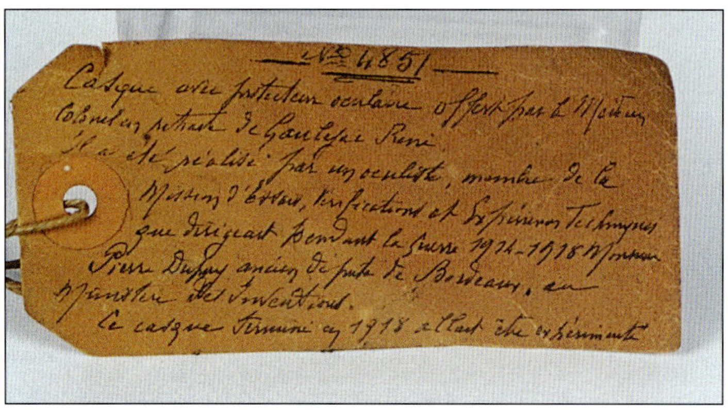

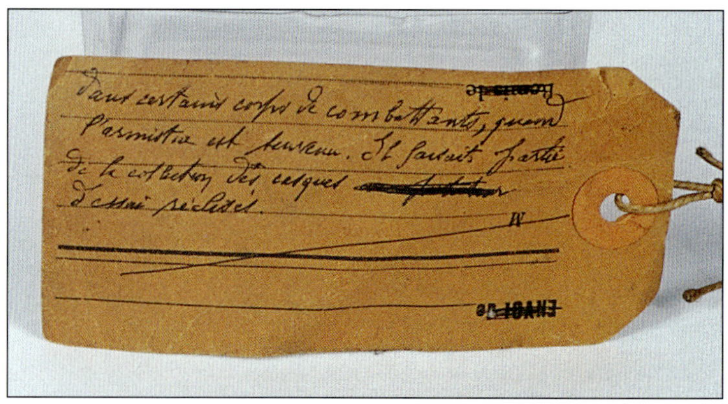

Museum label for previous Polack visored helmet which reads, "helmet with ocular protection donated by Dr. Rene de Grolejac (colonel retired). It was designed by an oculist, a member of the Prototype, Verifications and Technical Experiments Unit which was supervised during WW1 (1914-18) by Mr. Dupuy, former Bordeaux MP at the Ministry of Inventions. This helmet was completed in 1918 and was about to be tested with the Army when the war ended. It was part of a collection of prototype helmets".

Adrian Polack "Long-nose" variant.

Dunand helmet shell with riveted reinforcing to the brow. (Collection of Brana Radovic)

Dunand clip-on visor. Again, this was a way to convert issue Adrian helmets into visored helmets. The clips would hook onto the helmet rim then swivel down and lock when the visor was pulled down. (Collection of the Musée Militaire du Perigord, Perigueux, France)

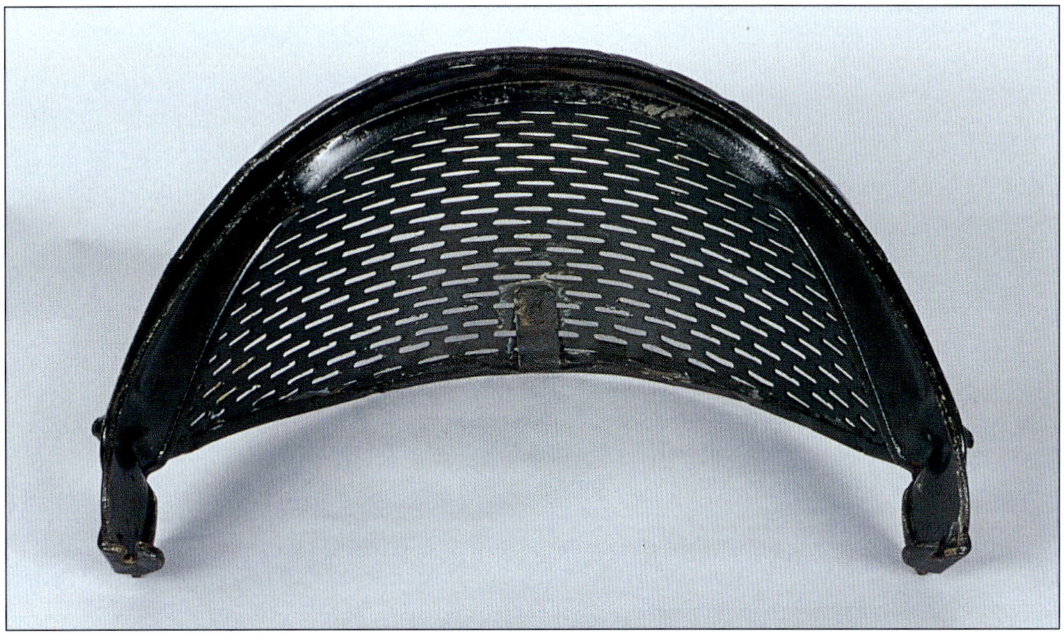

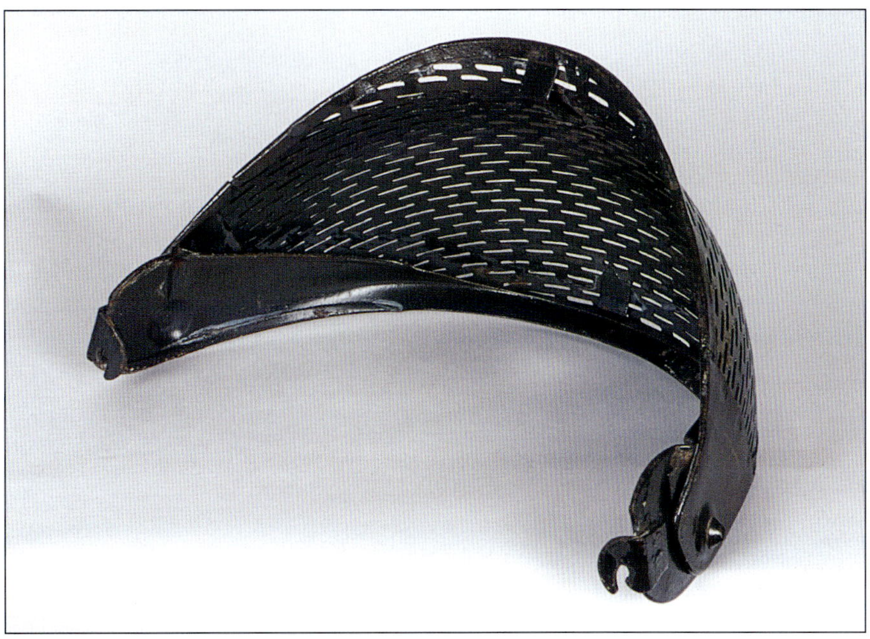

Officers' lightweight aluminium Adrian parade helmet. Painted blue with a standard steel/iron infantry badge and aluminium crest. (Collection of Ellwood von Seibold)

Officers' lightweight aluminium Adrian parade helmet. Painted blue with a non-standard aluminium infantry badge and aluminium crest. (Collection of Brana Radovic)

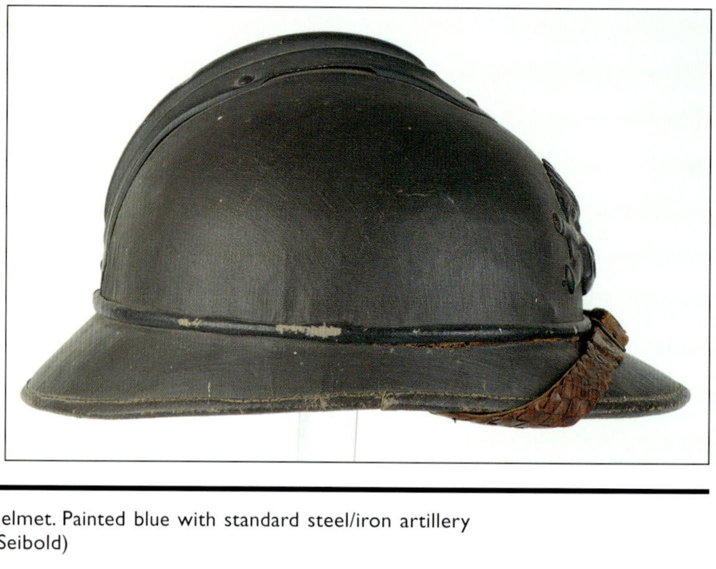

Officers' lightweight cork and cloth Adrian parade helmet. Painted blue with standard steel/iron artillery badge and aluminium crest. (Collection Ellwood von Seibold)

Top: The Reseda uniform. This was a complete helmet and uniform design produced in limited quantity in around 1912, but not adopted.

Bottom: A noted French general (General Temon) speaks to Australian troops. 7.5.18. Plane Corner, Bois de la Abbeys. AUST. CORPS. Note that the tall French officer in the centre is wearing a privately purchased Acier Trempe helmet with the Chasseurs' badge. IWM E(AUS)2153.

Soldier wearing an Acier Trempe helmet.

Postcard examples of Acier Trempe helmets.

Adrian Model 1915 helmet. Note this early helmet lacks the initials "RF" on the badge.

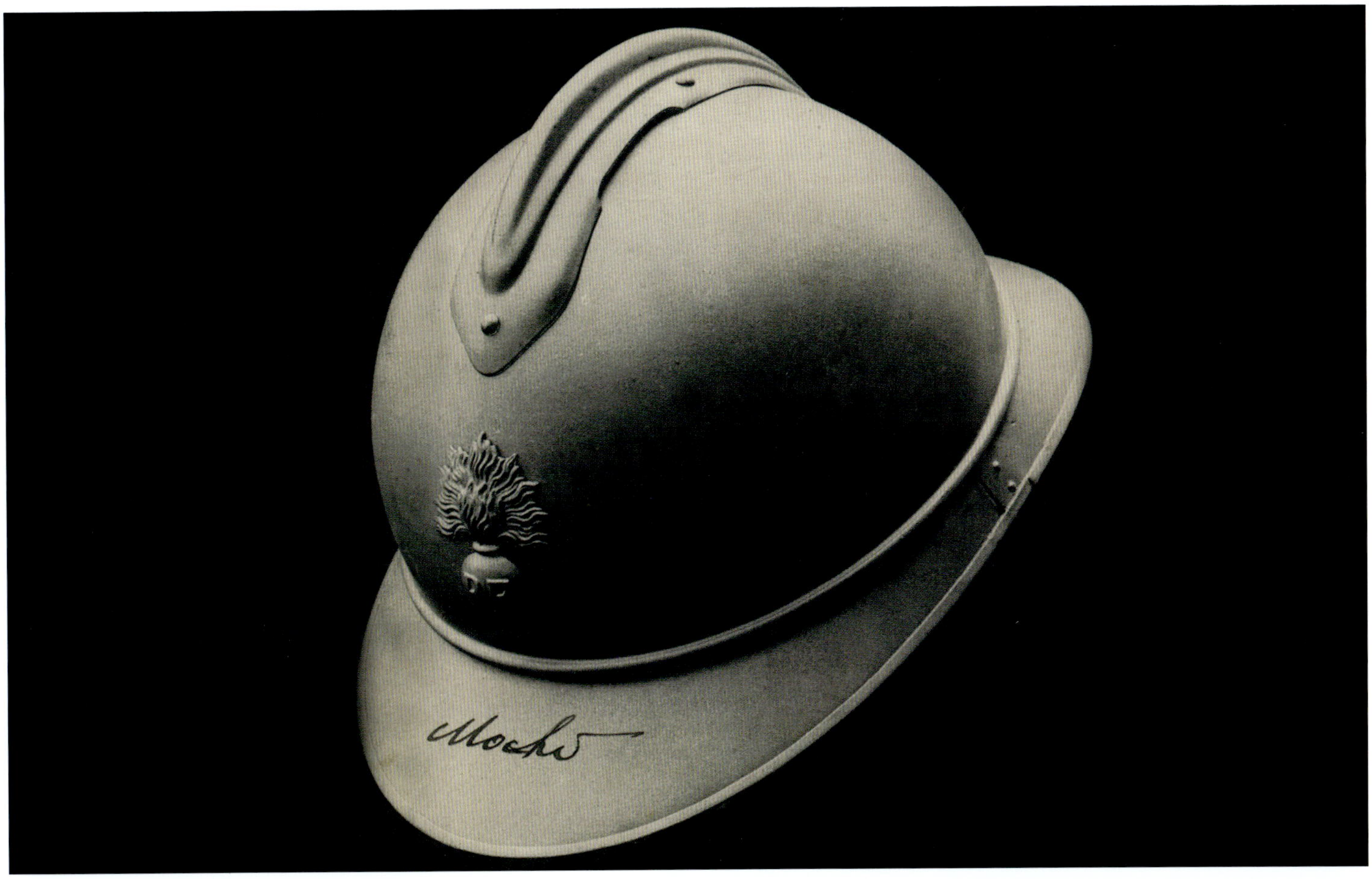

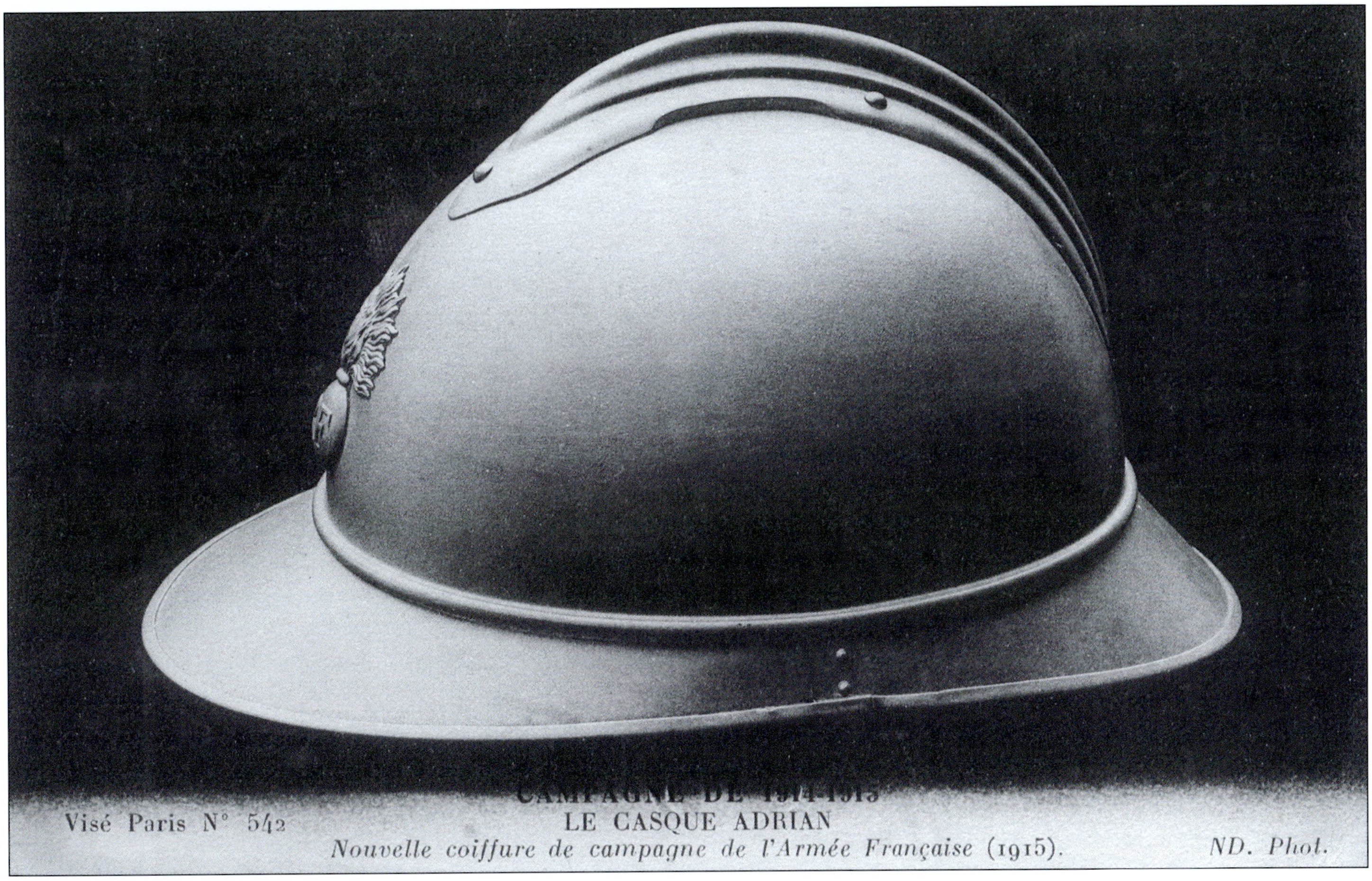

Postcards produced by helmet manufacturer la Societe des Phares Autoroche.

In front of La Ville aux Bois, Aisne. Bomb-thrower, 8th February 1916. IWM Q49095

■ P.2. Gas Mask with three compresses. La Renarde, Marne. IWM Q61055

Top: Working a 86mm pneumatic gun at the Servon Barricade, Marne. IWM Q79012

Bottom: A uniformed French artist sketching the damage in Amiens, 17 September 1918. Note the white-painted badges indicating Military police. IWM Q11345

Equipment of a Light Infantry private (chasseur a pied) Vincennes, 7.9.1916. IWM Q80290

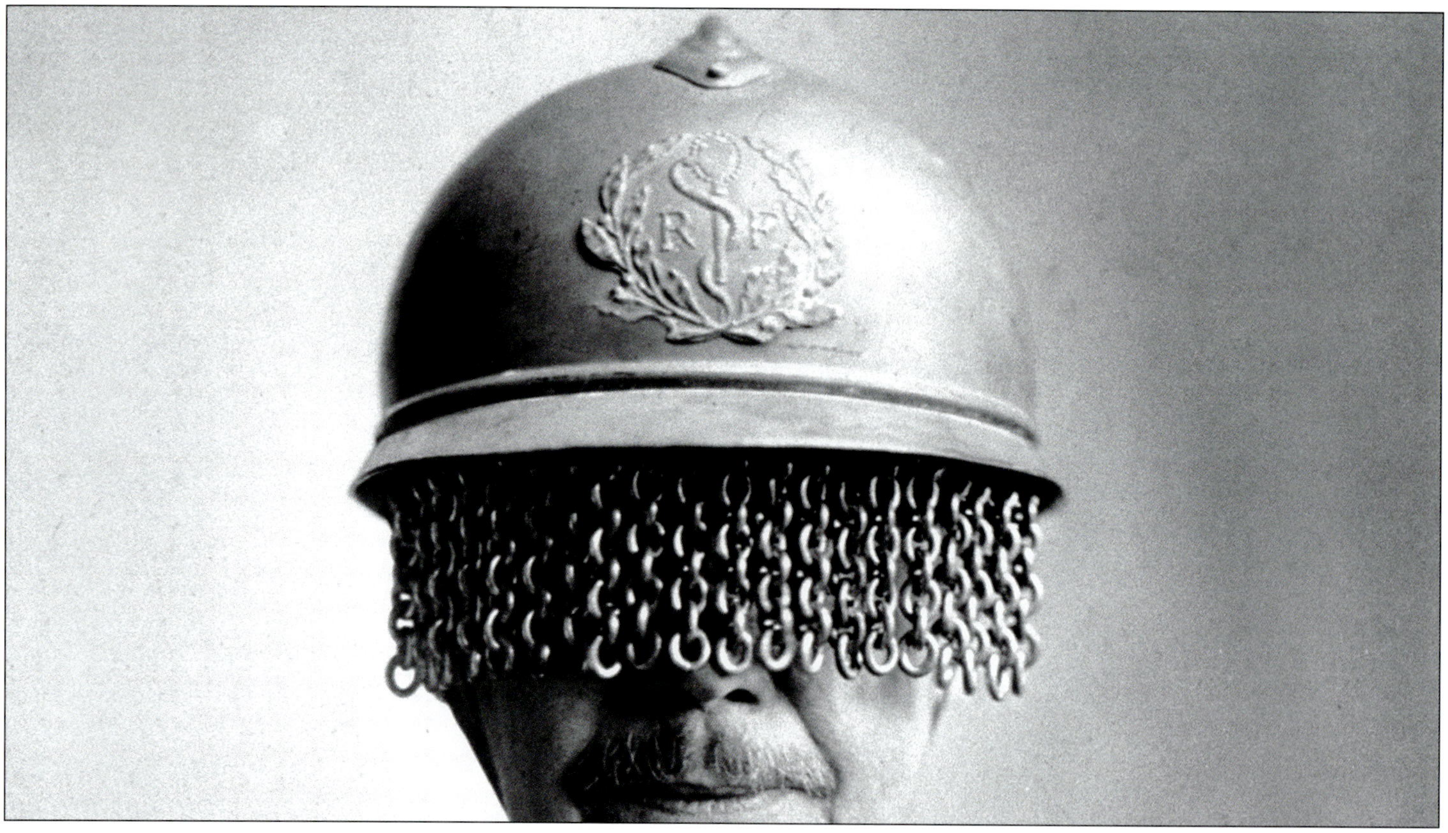

■ Top: Mail face protector worn by tank gunners: Ricicourt, 12th October 1918. IWM Q93249

■ Bottom: Mustachioed man wearing steel helmet with built-on chain screen to protect soldiers' eyes from fragments of shell, rock, etc. Manufactured by E.J. Codd Co., Baltimore, MD. Note the use of the Medical Corps badge.

Top: Medical troops badge on a propaganda postcard. Bottom: Administrative officer of the L'Intendeance. Note that his helmet badge is a unknown modification. Jean Marc Barris

Armoured observation post on the front line, Plateau de Vauclerc, Aisne. Oct 1917. Note the badge for Colonial Infantry and the Navy. IWM Q80288

Equipment of sapper of engineers. July, 1917. Vincennes. IWM Q50381

Battle of Cambrai, 1917. Civilians found in Masnieres about to leave the village in Official Photographer's car. 22nd November 1917. Note the triangular tactical patch to the front of the Adrian helmet cover. IWM Q3203

Army Clothing Department, Vanves. Regulation Infantry Uniform. October 1916. IWM Q80271

Top: Note Adrian Model 1915 helmet with cloth cover and Chasseurs' badge applied over the top. Stephen Chambers

Bottom: Senegalese entraining at the station at Frejus (Var), for the front. 20 June 1916. IWM Q78086

Top: French front line trench. Near Beaulne, Aisne. 4 July, 1916. IWM Q70050

Bottom: Adrian Model 1915 Infantry helmets. Stephen Chambers

NEW FRENCH WAR HELMET

The above photograph shows the new French trench helmet. With the added advantage of the visor down the soldier is protected from shrapnel and grenade fragments, which it is hoped will cut down the casualties to a small percentage Since the adoption of the steel helmet in use at the present time, the percentage of soldiers disabled from head wounds is very small.

Contemporary newspaper cutting showing the Dunand visor.

Top: Result of bullet tests on various types of helmets: Gondrecourt, 5th May 1918. Left to right: Dunand helmet with Polack visor, Dunand helmet with Dunand visor, Model 1917 Queen Elisabeth helmet, Dunand helmet without visor and another Dunand helmet with Polack visor. IWM MH31340

Bottom: The original caption to this photograph states that it was taken in Russia. The authors consider, however that the photograph showing a helmet work/repair shop was taken in France. Note the Chasseurs badges being fitted. Nik Cornish at Stavka Military Image Research

GERMANY

CHAPTER SEVEN

GERMANY

The German Army, on mobilisation in late July 1914, had in service a considerable variety of helmets. The best known of these are the Pickelhaube, or spiked helmet worn by the infantry, and the Kugelhelm, or ball-topped helmet, worn by the artillery. Generally speaking, jäger, aviation troops, and communication services wore the Shako. Mounted troops had their own helmets.

The authors recognise that the study of all the helmets in German service at the start of the Great War, and of those helmets developed from them, in any detail would merit a volume or volumes in its own right. A decision has therefore been reached to exclude from the photographic section helmets in existence when the war started and those helmets developed from them. However, the use of those helmets, to some extent at least, continued throughout the war. Some mention of them is appropriate. Where, in the following text, the Pickelhaube is referred to as the spiked helmet, that reference may be taken to include the Kugelhelm and Shako and other helmets in use at the start of the Great War.

It is hard to understand why the spiked helmet remained in service in 1914, let alone why that use was continued even after the introduction of a suitable steel helmet. The spiked helmet was of little protective value, but its use and manufacture continued. The French were able to quickly introduce one standard helmet, to be worn by all ranks for all purposes. The German experience was very different. There are a number of factors which, taken together, help to provide an explanation.

The first factor to bear in mind is the history of the use of the Pickelhaube. Adopted in 1842 by the Prussians, the design and introduction owed much to the personal involvement of King Frederick Wilhelm IV, who had been crowned King of Prussia in 1840. It was this helmet that evolved, by a series of changes, into the Model 1895 helmet worn in 1914.

Prussia was the pre-dominant of the northern German states. She brought the smaller states under her control. The use of the spiked helmet spread as Prussian influence and control spread. Following the Prussian victory in the Austro-Prussian War of 1866, the larger German states, including Saxony, Bavaria and Württemburg, became part of the North German Confederation. They were obliged, in the event of war, to support Prussia militarily. The Franco-Prussian War of 1870-71 produced a wave of nationalistic fervour uniting all the German states behind Prussia.

On the 18th January 1871, King Wilhelm I of Prussia was invested as German Emperor. A unified German Army was created. The spiked helmet became the standard headwear, including for those not already using it.

The short wars of 1866 against Austria and 1870 against France were fought under conditions that were very different to those of the Great War. Although a few modifications were made after those wars, the spiked helmet generally performed well. Had the conditions under which those two wars were fought been different, the deficiencies of the spiked helmet might have been exposed. However, this was not the case and the spiked helmet continued in service, despite being obsolescent.

In 1914, the German Army numbered 700,000 men. This number was increased to approximately 2,200,000 once reservists were mobilised. Over half of those serving were Prussian. Prussian officers dominated the senior ranks. The spiked helmet had become a symbol of Prussian tradition and German efficiency. There was considerable attachment to it, particularly among the officer classes, whose helmets had a distinctive design distinguishing them from the other ranks.

It is clear that there was resistance to the replacement of the spiked helmet. Many senior officers continued to use the spiked helmet, both in the front line and for ceremonial occasions, up to the end of the war and even though steel helmets were available. Indeed, the spiked helmet continued in use for ceremonial purposes even after the war ended.

Another factor in the continuing use of the spiked helmet was financial. The purchase of a spiked helmet by an officer represented a considerable outlay. The manufacture and supply of spiked helmets was an industry in its own right.

Thus, for reasons of tradition, class distinction and finance there was considerable resistance to the replacement of the spiked helmet, even after it became obvious that the helmet was unsuited to the conditions of modern warfare. This attachment to the old helmets undoubtedly cost lives.

The production of the spiked helmet, as it existed in 1914, required the use of cowhide to produce the helmet shell and front and rear brims. Prior to the start of the Great War, Germany obtained supplies of leather from Argentina. However, the British naval blockade quickly caused this supply to dwindle at a time when there was greatly increased demand due to the expansion of the armed forces. The demand for leather was not simply to manufacture helmets. Leather was used for all sorts of purposes, including military equipment such as belts and ammunition pouches. Additionally copper, used in the manufacture of the brass helmet fittings, was in short supply. Substitute, or ersatz, materials had to be identified if the manufacture of the spiked helmets was to continue. At a time when it might have been advantageous to concentrate on the introduction of a suitable steel helmet, considerable ingenuity and resources were devoted to the design and production of the spiked helmet in substitute materials.

The following information in respect of some of the helmets developed during the war may be of interest.

The Early Wartime Economy Pickelhaube

Faced with a grave shortage of helmets as the armed forces expanded, these were produced from old helmets, or using parts from old helmets, held in storage. As long as any helmet produced broadly conformed to the specification of the helmet in general use, there was no objection. Additionally, in order to conserve materials, the height of the spike was reduced on some helmets and some fittings were altered on some helmets. They were made thinner, or otherwise altered, to save metal. These helmets, in general terms of construction and materials, conformed to the standard designs and are not properly described as ersatz helmets.

The Model 1915 Pickelhaube, Kugelhelm and Shako

These followed a proposal made on the 25th August 1915 to produce standard helmets of the various types. The helmets were made of leather, but with steel fittings. The fittings were finished in a matte grey colour. The spike on the Pickelhaube was shorter and designed to be removable. A directive of the 21st September 1915 provided that the spike was to be detached when the helmet was worn in the front line and the cloth cover introduced was without provision for the spike.

These helmets are the most common types of Pickelhaube, Kugelhelm and Shako, being manufactured in millions between 1915 and 1917. They continued in use until the end of the war, particularly as far as artillery and rear echelon troops were concerned.

The Ersatz Pickelhaube

Some of these were derived from helmets designed many years before the Great War to be worn in the tropics. Others came about because of the shortages of materials that quickly developed once the war began. Helmets were manufactured from felt, which was the most commonly used ersatz material, tin, pressed steel, cork, papier-mache, vulcan fibre, aluminium, cloth and canvas. Helmets were constructed from more than one type of material. Old style helmets that had been withdrawn from service and helmets that had been damaged were cannibalised. Their parts, including the helmet shell, were used in the manufacture and repair of other helmets. Often, for instance, a front peak or neck guard was attached to a shell of different material.

The Ersatz Pressed Steel Pickelhaube

None of the ersatz helmets was introduced with the intention of providing the wearer with any degree of protection on the battlefield, save perhaps with one exception. The pressed steel pickelhaube, which came into service in early 1915 and was intended as a stop-gap measure pending the introduction of a proper steel helmet, was intended to provide a degree of protection which the earlier tin-plate spiked helmet did not.

When the ersatz steel pickelhaube was first introduced it had brass or cupro-aluminium fittings. By late 1915, painted steel fittings were used. The helmets were painted in grey or green and could thus be worn without a cloth cover, particularly once the early helmets had their fittings chemically dulled. Various manufacturing techniques were used. Some had a one-piece shell. Others had a lathe-turned bowl with the front and rear peaks being brazed on. Ventilation was provided by means of a circular hole in the apex of the helmet crown and five or six holes around the base of the spike. On early examples using brass or cupro-aluminium fittings the spike is non-detachable. On later helmets, the spike was detachable. The liner was of a one-piece construction in brown leather with seven tongues secured by a drawstring. The chinstrap was of black leather with fittings to match those on the helmet.

The helmet plate on the early helmets was secured by means of two loops attached to the helmet plate. Each of these passed through a hole in the helmet shell and was secured in place by means of a leather wedge on the inside of the helmet. Where the later grey painted steel fittings were used the helmet plate was secured by means of one loop and a peg.

The liner was secured by a split-pin at the front and rear of the helmet and by the chinstrap attachments on either side. These passed through the felt headband to which the leather liner was attached by stitching. The lower edges of the front and rear peaks were folded and there was no provision for front and rear rims nor for a rear spine. The helmets used by the various states making up the German Army vary in detail.

The Army Group Gaede Helmet

The first real attempt at providing a protective steel helmet, at the instigation of the Chief-of-Staff of the Army Group Gaede Lieutenant-Colonel Hesse, was in early 1915. The Army Group was engaged in operations in and around the Vosges Mountains in north-eastern France. In that rocky, mountainous region many soldiers suffered severe head wounds caused not only by actual projectiles but also by chips of rock. Lieutenant-Colonel Hesse was responsible for the development of a form of head protection. This was designed to protect the nose, forehead and top of the head.

On each side of the helmet there was an extension of the metal. Into this on both sides was cut a rectangular slot designed to take a leather strap which fastened by means of a buckle to the rear of the head. The chinstrap was attached on each side of the helmet by means of a slot cut in the metal with the strap passing through the slot. The metal part of the helmet was riveted to a skull-cap made of leather and cloth. The skull-cap was adjustable for size by means of a leather strap and buckle at the rear. Variations exist, including one type (illustrated in the photographic section) where there is no provision to take a strap to be secured across the back of the wearer's head.

The metal part of the helmet was made of steel varying between 5mm and 7mm in thickness. It weighed about 2.05Kg. Army Group Gaede produced the helmets at their artillery workshops at Mulhausen. Approximately 1,500 helmets were produced.

The helmet had obvious disadvantages. It was heavy and offered protection only to the front and, to some extent, the top of the head. No protection was provided to the rear and sides of the head. In those circumstances, once the German Model 1916 helmets were available, the Army Group Gaede helmets were withdrawn from service and the metal re-used.

German Facial Defence or Snipers' Face Plate/ Helmet

Little is known of this protection/helmet. It cannot be stated with certainty where they were made or when. It is probable they first saw service in 1915. There is mention that these helmets were manufactured by Becker AG Steelworks at Willich/Krefeld but this cannot be definitely confirmed. There were no markings on the example examined by the authors.

A report appeared in The Times Newspaper in about March 1916. This reported that one of the helmets was captured by the French in a German trench near Chaulnes. It was said to be three-eighths of an inch thick and to weigh slightly under fourteen pounds. The report included two sketches of the helmet and stated as follows:

> "As will be seen by the front section, it is shaped on the lower part of the right side in order that a man wearing it can use a rifle. In shape it resembles half of the headpiece of a suit of armour of the twelfth or thirteenth centuries. Although the eye-holes are only two very narrow slits, about five-eighths of an inch long and one-sixteenth of an inch wide, the wearer can see sufficiently to secure the range of the object to sight the rifle.
>
> The wearer puts the helmet on like a fencing mask, and is able to secure it firmly to the neck by the means of straps. The steel is shaped at various angles so as to deflect rifle and small machine-gun bullets. Tests which have been made upon the headpiece with bullets of small calibre show that the design is effective".

The helmet, it was said, was to be worn by sentries, snipers and those occupying exposed saps. Certainly, judging from the very small number of surviving examples, the helmets seem only to have been issued in a very limited number. Having said that, it may be that once the Model 1916 helmets were generally available, these helmets were scrapped and the metal re-used, as was the case with the Army Group Gaede helmets.

The authors have only been able to examine one example of this helmet and it is illustrated in the photographic section. The helmet itself was formed from one piece of steel approximately 8.5mm thick. The helmet measured at maximum approximately 33.5cm high and, 25.7cm wide. The rectangular eye-slits, viewed from the front, are very narrow indeed, measuring only 13.5mm wide on the right (as worn) and 15mm wide on the left, both being about 2mm high. The eye-slits widen towards the interior of the helmet to the extent that the interior measurements were 22mm wide and 5.5mm high. The helmet weighed 5.75Kg, although the padding would have added slightly to that weight.

The helmet had a metal framework fixed to the interior to take the fixing strap attachments and padding. The padding was absent on the example examined. The framework was fixed in place using five bolts. The exterior of each bolt was formed into a pointed cone-shape approximately 15mm wide at the base and 10mm high. It is perhaps worth mentioning that these cone-shaped points are identical to those found fitted to the low model of German loophole-plate. Each bolt was secured on the interior by means of a nut with a diameter of 15.5mm across the flat surfaces.

The framework to the interior was formed from two metal strips approximately 20mm wide and curved and bent to fit the interior surface. Fixed at the top by one bolt was a curved metal strip designed to go over the top of the wearer's head. This strip was formed from metal 22.5mm wide and 3mm thick.

The securing straps on each side were fixed at two points using brackets measuring 35mm wide and 39mm deep. One end of each bracket was slightly bent inwards and had a slot measuring 6mm by 20mm to take the straps. Each strap attachment was held in place by one of the securing bolts. The straps were formed of leather 24mm wide and 3mm thick, although the straps flared towards each end. The strap on the left (as worn) was constructed of two separate pieces of leather. These were joined at the centre by means of stitching and the leather base of the buckle attachment was placed over the join and again secured by stitching. The ends of the straps were permanently attached to the helmet again by stitching. The strap on the left appears to have had a pronged buckle at the centre and it is presumed the other strap, missing on the example examined, had holes punched to take the pronged buckle enabling the helmet to be secured at the rear of the wearer's head and to be adjusted to a comfortable fit.

The helmet examined was painted black inside and out. This included the interior metal fittings, although the paint there was mostly now absent, being replaced with light surface rust. The exterior of the helmet was over-painted a grey-green colour. It is presumed that the exterior of the helmet would have been painted to blend into the surroundings where the helmet was to be worn.

Overall, the helmet presents a somewhat hand-made appearance the metal, particularly the edges, being uneven in places. There is a tool-mark clearly visible in the corner of the cut-out section on the right.

The Models of German Steel Helmet

The Germans used four basic models of steel helmet during the Great War. The first helmet introduced was known as the Model 1916. Thereafter, the Model 1917 was introduced, the difference between the two models being that the earlier helmet had a liner band made of leather, whilst in the later model it was made of steel. Subsequently, the Model 1918 was introduced. The distinguishing features of that helmet were that the chinstrap was of different design and was attached directly to the liner band, rather than to lugs attached to the helmet shell as on the earlier helmets. Finally, a version of the Model 1918 was introduced on which the sides of the helmet adjacent to the wearer's ears were cut out and flared outwards. This version was generally referred to as the Model 1918 Cut-Out.

The differences between the various models, and some of the variations of the standard models that have been observed, are considered later and in the photographic section.

The Model 1916 Helmet

Design, Introduction and Production

Although the Army Group Gaede helmets were issued in limited numbers to that Army Group only, the problems that gave rise to the need for a suitable helmet were widespread. Large numbers of soldiers were sustaining head wounds. The "feldmutz" and other cloth headwear, and the leather or ersatz helmets in use, provided little or no protection. Communication took place between Friedrick Schwerd, who was a professor at the Technical Institute in Hannover and also a captain in the Landwehr, and Professor August Bier, who was the Naval Doctor-General and consulting surgeon to the XVIII Army Corps. The dialogue was prompted not only because of the alarming number of head wounds, many of them caused by small projectiles no bigger than a pea, but because of the progress the French were reportedly making with the development of their helmet.

Professor Bier wrote to the medical staff of the 2nd Army on the 15th August 1915. There followed a detailed consideration into the type and design of helmet required. This included investigation and testing of the type of steel to be used. Following this, on the 14th December 1915, the Prussian War Ministry wrote to the Bavarian War Ministry announcing that preliminary trials had been completed to the extent that 30,000 helmets were to be produced and issued for field trial by the end of January 1916. In fact, it appears that the first helmets were received for field testing in December 1915. The first 30,000 helmets were produced by Eisenhuttenwerk at Thale and were issued by the end of January 1916. Following successful use, the helmet was authorised for general issue some time in February 1916.

With production getting under way, troops fighting at Verdun and on the Somme, where the Allies launched their offensive on the 1st July 1916, were given priority for issue of the new helmets. By the end of July 1916, some 261,000 helmets had been distributed. As the helmets were issued, so the spiked helmets with which the troops had previously been equipped were returned for use by troops in training and in reserve. The process of equipping the troops with the new helmets was by no means instantaneous. Photographs exist where, within the same unit, some soldiers have the new helmet and others continue to wear the spiked helmet.

By the 11th November 1918, approximately 7.5 million steel helmets had been produced together with about 50,000 frontal (brow) plates. Helmets were also supplied to Austro-Hungary, Bulgaria and Turkey.

By way of general description, the helmet was of one-piece construction. Chromium-nickel steel approximately 1mm thick was used for the helmet shell. The helmets were formed from pre-cut circular sheets of steel. There were seven stamping steps with the metal being subjected to four separate annealing, or tempering, processes. The helmet had a visor which, when the helmet was worn, was intended to be level with the eyes. At each side of the helmet, level with the wearer's temples, there was a step downwards leading to the side and then rear apron of the helmet. When worn, the bottom of the apron was intended to be horizontal and the width of two fingers from the wearer's head. Approximately 5mm of the helmet rim was folded inwards and pressed flat to produce a smooth finish.

Ventilation was provided by means of two holes, one on each side of the helmet, above the wearer's temples. Into each hole was fixed a metal lug. The lugs were hollow to allow air to pass through. Additionally, the lugs served to protect the ventilation holes and as a means of attaching the brow plate to the helmet. The lugs were attached to the helmet by folding the inside edge of the lug over a washer on the inside of the helmet shell. The washer was 22mm wide and provided a firm surface against which to fold the end of the lug.

Three holes were punched around the circumference of the helmet, two at the front sides and one at the centre rear. The holes were punched in the helmet shell just above the point where the apron commenced to flare outwards. The split pins used to secure the liner in place passed through the holes and the liner and were then folded flat against the inside of the liner band.

On each side of the helmet on the Models 1916 and 1917 a hole was punched to take the lugs to which the chinstrap was attached. Each chinstrap attachment lug was made up of three components. Firstly, there was a securing rivet. Usually a dome headed rivet was used, although rivets with rounded sides and a flat top to the head have been observed. The rivets were inserted from the exterior of the helmet through the pre-cut holes. The rivet passed through two washers and was then fixed by hammering. The first washer measured 14mm wide and 5mm deep. The second washer measured only 2mm in depth. It had the same diameter as the first washer but had a triangular projection. This was fixed so as to point upwards and towards the rear of the helmet. This triangular projection served to retain the chinstrap attachment in place.

The Manufacturers of German Helmets

The following is a list of the most important manufacturers of German helmets during the Great War with, where known, details of their location and manufacturing code(s):

Manufacturer's	Name	Location Code(s)
F.C. Bellinger		BF, FB
Eisenhuttenwerk Thale A.G.	Thale/Hartz Thuringia	ET
Gebruder Gnuchtel A.G.	Lauter/Sachsen	G
Sachsen		GL
Gebruder Bing A.G.	Nurnberg	GBN, B
Stahlwerk Richard Lindenberg A.G.	Remscheid-Hasten	LI, L, K
Vereinigte Deutsche Nickelwerke	Schwerte/Westphalen	NJ
Vereinigte Deutsche Nickelwerke	Schwerte/Rheinland	DN
Firma F.W. Quist	Esslingen	Q, QL
Sachische Emailles u. Paruschowitz	Oberschlesien	Si
Siemens u. Halske A.G.	Siemensstadt bei Berlin	SH
C. Thiele & Sohne	Lubeck	TJ, J, T, FT
Herman Weissenberger & Co	Stuttgart/Canstatt	W
Stahlwerk Rockling A.G.	Volkingen	
Bismarkhutte	Oberschlesien	Bi
Stahlwerk	Krefeld	B
Korting & Mathiesen	Leutzch bei Leipzig	
J & H Kerkmann	Ahlen/Westphalen	

It is not known what percentage of the total production was contributed by each manufacturer, nor for what periods, in each case, the manufacturer produced helmets. What can be said, based on a study of surviving helmets, is that Eisenhuttenwerk at Thale (ET) was probably the pre-dominant manufacturer and that they produced helmets throughout the war.

The manufacturer's stamp, by way of an identifying letter or letters, was almost invariably stamped inside the left-hand side of the helmet shell opposite the wearer's temple. Some of the manufacturers are, it is hoped, correctly identified in the above table. The identity of other manufacturers and some of the markings is unfortunately lost or not known at the present.

Helmet Sizes

The helmet shells were manufactured in five sizes. There has been mention of a sixth size, size 70, but the authors have never seen such a helmet. The helmet size was almost invariably stamped into the metal in the left-hand side of the helmet, immediately following the manufacturer's initial(s). In addition, the size was sometimes stamped in ink inside the rear apron of the helmet shell, usually near the centre. It is impossible to say whether all helmets were thus marked. There is on many helmets no visible ink stamp but whether this is because the helmets in question were never marked or because the stamp has faded or has been erased cannot be said. In addition, an ink acceptance stamp "AK" appears on some helmets, usually on the rear apron of the shell.

Some helmets have, in addition, been observed with the size stamped into the leather or metal of the liner band, usually on the left, near or at the front.

The sizes were as follows:

Size 60 to fit head sizes 50mm to 52mm
Size 62 to fit head sizes 52.5mm to 54mm
Size 64 to fit head sizes 54.5mm to 56mm
Size 66 to fit head sizes 56.5mm to 58mm
Size 68 to fit head sizes 58.5mm to 60mm

The following table gives details of the average helmet dimensions together with the percentage of the particular helmet sizes compared with a total of fifty-five helmets surveyed. It is to be stressed that the figures given are an average of the helmets examined. Helmets will be found where the measurements vary significantly. In this respect it is worth noting that the front to rear measurements of helmets of the same size manufactured by the same manufacturer varied by up to 1cm. Damaged helmets were excluded from the survey.

As to the percentage figures given for any size compared with the total number of helmets examined, it should be noted that the helmets were collected without reference to the size of any individual helmet. The figures given may thus give some indication, by way of general guidance, of the percentage of any given size compared with the total of all helmets manufactured. However, the actual production figures for the percentage of helmets of any given size are not known.

Helmet Dimensions

Size	Front to Rear	Side to Side	% of total
Size 60	298mm	231mm	4%
Size 62	302mm	226mm	16%
Size 64	311mm	235mm	40%
Size 66	317mm	237mm	35%
Size 68	323mm	243mm	5%

The only measurement that does not fit into a pattern of gradually increasing dimensions is the width of the size 62 helmets. The measurement given is an average of nine helmets with the narrowest measuring 220mm and the widest 233mm. The two size 60 helmets in the survey were both manufactured by a manufacturer using the initials SH with the H stamped over the S. This is thought to be Siemens u. Halske AG.

In addition to the dimensions of the various sizes, there was a difference in the lugs protecting the ventilation holes and serving as fixing points for the brow plate. The reason for this was that, although the helmet shells were produced in a number of sizes, there was only one size of brow plate, designed to fit all helmets. It was therefore necessary to add, in respect of the three smaller sizes of helmet, a step at the base of the lugs to take up the space created by the reduction in size of the helmet shell and thus preventing the brow plate from being too large. The size 60 helmet had a 7mm step, the size 62 a 5mm step and the size 64 a 3mm step. The two larger sizes had no step. All the steps were approximately 18mm wide.

The outer part of the lugs on the three smaller sizes, and the lugs on the larger two sizes, protruded by approximately 8mm and were 8.5mm wide. The ends of the lugs were normally cut square, although on some helmets, depending on the manufacturer, they were rounded to a greater or lesser extent.

It has been observed that the ventilation holes on some helmets have been deliberately blocked. Often the material used to block the holes was, it appears, chewed paper. It is noted that a report dated the 23rd March 1916 in respect of the original trial helmets, in answer to the question:

> *"Is the ventilation arrangement to be viewed as purposeful, or was an unpleasant feeling of coldness evident even after the plugging of the vent holes and the covering of the head with a cloth".*

Concluded:

> *"Without a cloth the ventilation was too strong and caused a feeling of coldness".*

It appears, in some cases, that the blocking of the ventilation holes was in order to prevent the draught. However, there is also reference to blocking the ventilation holes to reduce noise (see comments under Model 1918 Cut Out) and in the circumstances it is impossible to say why the ventilation holes on any given helmet were blocked.

The Liners

Before commenting on the standard liners, it is perhaps worth mentioning that variations from the standard liners may occasionally be found. The original design anticipated a liner with three tongues on each liner pad, rather than two tongues as found on the standard liner. In the photographic section there are illustrated two helmets with liners that are far from standard.

In addition to variations in design, variations in the materials used may be encountered. For instance, one helmet observed, a "square dip" helmet marked ET64, has the liner pad pockets formed from leather instead of cloth. Many different types and patterns of material were used to form the liner pad pockets and to cover the pads inserted into them. A Model 1918 helmet has been observed where the pads are made of cloth with the padding stitched in, rather than there being provision for the insertion of a separate pad. This appears to be an economy measure rather than anything else.

These variations, and no doubt others, must be borne in mind. It must certainly not be assumed that a helmet that differs from the "standard" design is not perfectly genuine.

The Model 1916 Liner

The liner consisted of a leather headband secured to the helmet by means of three split-pins. Each split-pin passed through a 7mm pre-cut circular hole in the helmet shell. The holes were cut two at the front on each side below the level of the ventilation lugs and approximately 45mm away from them, the third being at the rear centre of the helmet shell. Each split-pin passed through the helmet shell from the exterior and then through a hole cut in the headband. The ends were then folded flush with the headband, thus securing it in place. The heads of the split-pins were approximately 10mm wide. The heads of the two at the front were only 1mm to 2mm deep, whilst that at the rear was 3mm to 4mm deep. The extra deep head of the split pin at the rear was designed to prevent the strap securing the brow plate from riding up.

The leather headband was usually 45mm deep and was normally constructed of one thickness of cowhide. However, there appears to be a considerable variation from helmet to helmet in the thickness of leather used, with some liner headbands constructed from a double thickness of leather. Where a single thickness of leather was used, the liner pads were secured by a single row of stitching passing entirely around the circumference of the headband. Where a double thickness of leather was used to construct the headband, there were two additional rows of stitching serving to hold the two bands together. The leather headband had a U-shaped section cut from it on each side thus ensuring that the headband did not obstruct the ventilation holes. The ends of the leather headband were joined at the rear of the helmet, just to the right or left of the rear split-pin, by a row of six stitches. Twine, often of double thickness, was used for this purpose.

There were three liner pads. They were attached to the exterior bottom of the headband by stitching. The liner pads were then folded over the bottom of the headband to bring them to the interior of the helmet. Each liner pad was rectangular shaped and divided into two tongues. The tongues extended towards the apex of the helmet. Each had a small hole near the end. Through this a drawstring, used to adjust the depth to which the helmet sat on the wearer's head, was threaded.

Each of the three liner pads had a pocket to the interior between the liner pad and the headband. Each of these pockets was designed to take a liner pad cushion. The liner pad pockets and cushions were normally formed from cream coloured, linen-type material. However, many different types of material have been observed, including quite brightly coloured and patterned cloth. Each cushion was held in place by tying across the mouth of the pocket, usually by means of a bow, two pieces of cotton tape. A number of different colours were used for this tape including white, cream, pink and black. The ends of the cotton tape were respectively attached by stitching to the base of the liner pad and to the pocket itself. The preferred filling for the liner pad cushions was horsehair. However, other fillings including straw and waste material were used.

The headbands and liner pads were, on the Model 1916 liners, almost invariably made from brown leather. The leather headband may be marked with a stores stamp giving details of the army group and the year. Very occasionally, manufacturers marked the leather headband, usually on the side facing the helmet shell, with their name and place of manufacture. The temptation to remove the liner to see if this feature is present is to be resisted. Invariably damage will be caused and the helmet devalued.

Occasionally, the liner pads were marked with the manufacturer's details and year of manufacture. This marking may be found on the part of the liner pad that goes against the wearer,s head, close to where the liner pad divides into two tongues. For instance, in the collection of one of the authors is a Model 1917 helmet where all of the liner pads are in brown leather and are marked "1916. Otto Koch & Co. Berlin S.O."

The Chinstrap

The standard chinstrap consisted of a single strip of leather measuring approximately 15mm wide and 615mm long. It must be stressed that these measurements may vary considerably. Indeed, chinstraps as short as 480mm have been observed. It may be that the shorter chinstraps have been removed from spiked helmets and re-used. Variations in width also exist.

At each end of the chinstrap was a buckle, without prong. The buckles measured approximately 22mm wide and 26mm long. Again, these measurements vary. The chinstrap passed through the buckle. A short length at each end of the chinstrap was then passed through the centre strut of the buckle and folded back. Each end was secured by a single row of stitching, using twine. The buckles served as a means of adjustment of the length of the chinstrap.

The means of attachment to the helmet was the same system as found on the spiked helmets. The attachments on each end of the chinstrap consisted of a piece of rigid metal. At one end was a circular hole with a V-shaped piece cut out of it. This was designed to fit over the chinstrap attachment lug on either side of the helmet. At the other end of the piece of rigid metal was a rectangular slot through which the chinstrap passed. The early chinstrap attachments were made of brass. Later fittings were of zinc-plated steel.

The chinstrap was removable by simply revolving each chinstrap attachment on the helmet lug until the V-shape on the lug lined up with the V-shape cut out on the chinstrap attachment. The chinstrap attachment could then be lifted off the lug. When fitted to the helmet but not in use, the chinstrap could be pulled up over the helmet visor.

It seems clear that the method of chinstrap attachment was inadequate. On surviving helmets the chinstrap is often missing. Although some chinstraps may have broken and were not replaced, it seems likely that many simply dropped off and were lost. On some helmets the chinstrap lugs have been deliberately distorted by hammering thus preventing the removal of the chinstrap. It is not known whether this unofficial modification was necessary to prevent the chinstrap falling off accidentally, or whether it was to prevent removal of the chinstrap by another individual who needed a replacement for his lost chinstrap.

Chinstraps are usually in brown leather, although some are black. It may be that those in black are chinstraps taken from spiked helmets.

Paint Finishes

The standard paint finish was a grey-green colour described as fieldgrey. Once the helmet shells had been pressed, and the chinstrap and ventilation lugs fitted, the helmets were dipped in paint and then fired at a temperature of 120 degrees centigrade. There was considerable variation in the paint finishes. Some helmets had a matte finish, whereas on others the finish was reflective. The shade and thickness of the paint used also varied. On some helmets the paint is so thick that it has run in places. It is clear that the paint was generally found to be reflective in damp or sunny weather or in the moonlight. To remedy this, mud was applied to the exterior of the helmet shell. Subsequently, cloth covers were used or a camouflage finish applied.

There were two basic types of camouflage finish. The first was in accordance with the order of the Chief of the General Staff of Field Army II (Ludendorff) No: 91366 of the 7th July 1918. That provided for the painting of all models of helmet (as well as artillery pieces, machine guns etc.) in a variety of colours. It was said that green, yellow, ochre, rust and brown were suitable colours for that time of year. It appears that a three-coloured surface was anticipated, with one of the three colours matching the basic colour of the area of operations. The colours were to be separated by finger-width black lines. To the front of the helmet, no more than four of the areas separated by the black lines were to be visible. The order contemplated the use of fine sand to be sprinkled on to the wet paint surface to provide a matte finish. It was anticipated that the colour scheme would be varied with the changing of the seasons.

Some helmets were possibly given a painted camouflage finish complying with the order before leaving the factory. These helmets may today generally be distinguished by the quality of the painting, with the paint being evenly applied with broad, accurately painted, black separation lines. In other cases, a detail of men in a particular unit carried out the work. Otherwise, an individual painted his helmet. Thus, there was considerable variety as far as quality of painting and finish and the design and the colours of paint used are concerned. The camouflaging of helmets almost certainly commenced before July 1918 and in any event, many of the finished helmets did not comply with the order.

The second type of camouflage finish was known as "flocking". This finish was referred to in a directive of the 1st August 1918 whereby it was proposed that 100,000 of the new cut out helmets would be finished with a matte, felt-like finish using wool dust. This finish would not shine and would replace the cloth covers and camouflage paint finishes in use. Soil, it was said, could be smeared onto this new finish to better match the helmet to the surroundings. The soil would adhere better to this new finish than to the old painted surface. The soil could be removed by light brushing or washing. It is not known by what method the wool dust was fixed to the helmet surface. The finish was only applied to some of the Model 1918 cut-out helmets, many being painted in standard fieldgrey paint.

Helmet Covers

It has been said that 800,000 helmet covers were issued during the war. However, details as to any official pattern are not available. Although contemporary photographs show covers in use, the extent of that use is not known. The first covers, issued in January 1917, appear to have been in white to provide camouflage in snow. However, there is later reference to enquiries in February 1917 as to the issue of fieldgrey covers. The covers observed range from white to beige and fieldgrey to brown. Material used was mostly cloth. However, certainly hessian was also used. On some covers there was no provision for the ventilation lugs. Other covers had appropriate holes through which the lugs protruded. In some cases the holes were reinforced with leather or by additional stitching.

Attachment to the helmet was sometimes by drawstring. In other cases, it was by means of metal hooks attached to the base of the cover and which hooked over the helmet rim. This was the system often used on the spiked helmets.

Original Great War covers are scarce. The existence of reproductions has served to confuse the position since sometimes the reproduction is indistinguishable from the original.

Insignia

Generally speaking, no insignia were worn by German troops on steel helmets, or steel helmet covers, during the Great War. One exception to this, confirmed by contemporary photographic evidence, was the Hohenzollern crest worn by the 1st Foot Guards Regiment and the 1st Guards Signal Battalion. The crest, a shield divided into two white and two black sections, was painted on the centre of the helmet on the left, sometimes with a unit number painted to the rear of the helmet.

Other painted symbols, such as a stylised machine gun or state shield painted on the left side of the helmet have been reported or observed. However, there appears to be no official order or sanction in respect of the painting of helmets with symbols or insignia during the Great War.

The Brow Plate

The brow or frontal plate (stirnpanzer), was introduced at the same time as the Model 1916 helmet. It appears that the initial intention was to issue one brow plate with every twenty helmets. The number actually produced is unknown, although it has been suggested that the number was about 50,000.

There was only one size of brow plate, designed to fit all the helmet sizes. The brow plate was made from the same steel as used to make the helmets and was 5mm to 6mm thick. At the base, which was somewhat flared, the brow plate measured between 256mm and 258mm wide. Across the slots cut in each side to accommodate the ventilation lugs the brow plate measured approximately 202mm. On each side, between 13mm and 15mm below the slots that fitted over the ventilation lugs, there were rectangular slots cut in the metal to attach the securing straps. These slots measured between 17mm and 20mm deep by about 7mm wide.

The securing straps were attached by two rivets on each side. One strap, on the left, had a pronged buckle at the end. The strap on the right had a number of holes, far more than was necessary on the examples examined, to take the buckle. Why there were so many holes is unknown, but it may be that the straps were originally intended for some other purpose. On several examples examined the end of the right hand strap is marked with the year (1916) and manufacturer. When worn, the brow plate was supported on each side by the ventilation lugs and rested on the helmet brim. It was secured to the helmet by buckling the strap at the rear. The large head on the rear liner headband securing split-pin prevented the strap from riding up.

The brow plate protected the wearer's forehead and temples. The top and sides of the head were also, to some extent, protected given that the brow plate extended outward from the helmet. It was said that the brow plate would protect the wearer from a rifle bullet fired from 50 metres or more away.

The weight of the brow plate was about 2kg. It could thus be worn for short periods only.

The Model 1917 Helmet

In May 1917 the standard helmet was modified by the substitution of the earlier leather headband with one of pressed steel. The reason for this was the shortage of leather that had developed. It should not be assumed that the introduction of this modification, creating as it did the Model 1917 helmet, was instantaneous. Stocks of existing liners were, of course, used up. It is thus perfectly possible to find, for example, a Model 1917 helmet with liner pads dated 1916.

The new steel headband was 28mm deep. As with the Model 1916, it was held in place by three split-pins. The lower edge of the headband was folded outwards, that is against the helmet shell. At the intervals where the liner pads were to be fitted, the fold on the bottom edge was pressed slightly flatter than elsewhere thus crimping the liner pads in place. Beneath the liner pads, the liner headband was pierced by between 32 and 34 small holes. The original intention appears to be that the liner pads would be held in place by stitching with the small holes to be used for that purpose. A few helmets have been observed where stitching has been employed, but the majority of helmets were not stitched, the crimping being adequate to secure the pads.

There were no other differences between the Models 1916 and 1917. However, it is perhaps worth mentioning that white or cream coloured leather was routinely used for the liner pads with the Model 1917 helmets. This was not a feature seen on the Model 1916 helmets. The white or cream leather was rabbit skin and its use appears to have been quite widespread from 1917 onwards.

The Model 1918 Helmet

This helmet was virtually the same as the Model 1917 save for the design of the chinstrap and the chinstrap attachment. In the Model 1918 helmet, the chinstrap was attached directly to the liner headband rather than by means of lugs attached to the helmet shell. The absence of the chinstrap lug on each side of the helmet interior and of the rivet heads on the exterior make it easy to distinguish the helmets at a glance.

The chinstrap attachment was a rectangular wire loop. The loops measured 22mm wide by 15mm deep. The rectangular loops were attached to the steel headband by means of folded rectangular pieces of metal. These were folded around the rectangular loops and then riveted to the inside of the headband by means of a rivet passing from the interior of the helmet. The rivets had rounded heads and a washer was placed between each rivet head and the headband. The washer prevented the rivet passing through the headband and added strength and rigidity to the fixing.

The chinstrap was in two parts. The left-hand part was passed through the rectangular loop, the end being secured by two small rivets. This section of the chinstrap had a metal loop attached to the end with a leather tongue beneath it to protect the wearer's chin from the clip on the other part of the chinstrap. The leather tongue was usually marked with details of the manufacturer and 1918, being the year of manufacture. The right hand section of the chinstrap passed through the rectangular chinstrap attachment. One end was attached to the adjustment buckle. The other end was attached to a spring-loaded fastening clip. The clip was hooked to the loop on the other section of the chinstrap and the spring-loaded section prevented the clip coming undone accidentally.

The Model 1918 Cut Out

This helmet was identical in all respects to the Model 1918 save that on each side, adjacent to the wearer's ears, a section of the helmet measuring approximately 25mm deep by 90mm long was removed. The inward folding of the rim was continued around the cut out section which was flared outwards, thus creating a smooth, flowing contour to the side of the helmet.

It was hoped, by cutting out a section of the helmet, that the wearer would be able to hear better. In addition, the cut out section fitted the shape of the field telephone then in use, thus enabling the telephone operator to use the telephone without removing his helmet.

A memorandum dated the 1st August 1918, dealt with the issue of the Model 1918 cut out helmets, including the flocking finish referred to earlier. That memorandum, it is interesting to note, referred to the stuffing of one or both of the ventilation lugs with paper plugs to reduce noise. The memorandum required a response, by way of completion of a questionnaire, including a comparison with the earlier helmets, by the 1st December 1918.

It is not known how many Model 1918 cut out helmets were produced. Certainly the number was relatively modest. The only cut out helmets observed so far have been marked ET 64.

Variations of the Standard Helmets

Although little is known about the variations of the standard helmets that have been observed, it is appropriate to mention at least some of them. They are as follows.

The Square Dip Variation

This was a variation of the standard Model 1916 helmet in which the dip between the visor and the apron was far sharper than usual. This sharp angle was, it is said, a design fault by the Juncker Co. in the specification of the original helmet. The fault was, in effect, caused by bringing the rear apron too far forwards before it turned upwards. This created a sharp angle and caused excessive stress in the metal, resulting in a large and unacceptable number of failures when the helmet shells were pressed.

The design was changed, although the original square dip helmets were issued and used. The number produced must be relatively small, as surviving helmets are scarce. However, this may be explained to some extent by the fact that they were the first helmets issued and, therefore, the survival rate may be lower than later helmets. Some surviving examples have been reported as having a crack or cracks at, or close to, the square dip. This was presumably caused by the stress placed upon the metal.

Some features noted on square dip helmets are as follows:

- The sharp step between the apron and the visor on each side.
- All were made by Eisenhuttenwerk Thale A.G. and are stamped E.T. Most are size 64. However, at least one example of a size 66 exists and is illustrated in the photographic section.
- Where a standard liner was fitted, all have a liner headband that was made of thinner leather than usual.
- The liner headband was almost always stamped with the size in the front or front left.

The Reinforced Helmet

This was a helmet with an extra thickness of steel riveted to the front. The reinforcing plate was fitted over the front visor and extended upwards and rearwards to the ventilation lugs. The plate, it appears, was fixed in place using approximately 12 rivets.

The authors have seen an example but, unfortunately, have not had the opportunity to examine it in detail.

The Helmet with External Chinstrap Lugs

This was a highly unusual variation although several examples have been noted. The chinstrap lugs were attached to the outside of the helmet, in effect, the wrong way round. This was, in fact, the position of the lugs on the spiked helmets. However, whether the placing of the lugs on the steel helmet was an experiment or an error is, at present, unknown.

The Helmet with Dual Chinstrap Attachments

This was a helmet having the standard Model 1916/17 chinstrap lugs. However, it had additional chinstrap attachments fitted to the interior on each side of the helmet shell approximately 45mm behind and 20mm above the position of ventilation lugs. The additional attachments were approximately 100mm above and in line with the original lugs.

The additional attachments comprised a piece of bent steel with a square base and rounded top. This was riveted to the helmet shell. The attachments had rectangular slot cut in them to take the chinstrap.

The liner appears at first glance to be a standard Model 1916 liner with leather headband. However, closer examination reveals that the liner was made up from two sections. The sections were joined on the right hand side using four stitches of double twine. This joint was where the headband had a U-shaped cut out section to allow for ventilation. In addition, the liner headband was joined at the rear with the usual six stitches. The additional section of liner headband was required because, on each side below and slightly to the front of the additional chinstrap attachment points, the liner headband was bowed inwards. This bowing on each side allowed the chinstrap to pass between the helmet shell and the liner headband.

The liner headband was fixed, not by means of the usual split-pins, but rather by means of three dome-headed rivets inserted from the exterior of the helmet. A steel washer was used on the interior of the liner headband and the rivets were hammered in place against these, thus securing the headband in place.

The helmet shell of the example examined was standard and marked in the usual place Q 66. The liner headband has an ink stores stamp. This appears to read BAX III 1916. It is thought these helmets may represent an early attempt to improve chinstrap attachment. Alternatively, it is possible this modification was introduced for a particular purpose where the helmet was required to be more securely attached to the head than usual, such as if the helmet was being used in an aircraft. In this respect, the additional chinstrap attachments appear to have been fitted after the hèlmet shell was painted. The fittings are not perhaps of factory quality and this may indicate non-factory modification.

The Full Visor Variation

This was a variation in which the visor sloped forwards from the top of the helmet at a increased angle. This may have been an experiment to allow greater clearance between the wearer's forehead and the helmet to allow the more convenient use of a gas mask. Alternatively, the increased angle at the front of the helmet may have been an experiment to ascertain if projectiles would be better deflected. Very little is known of these helmets.

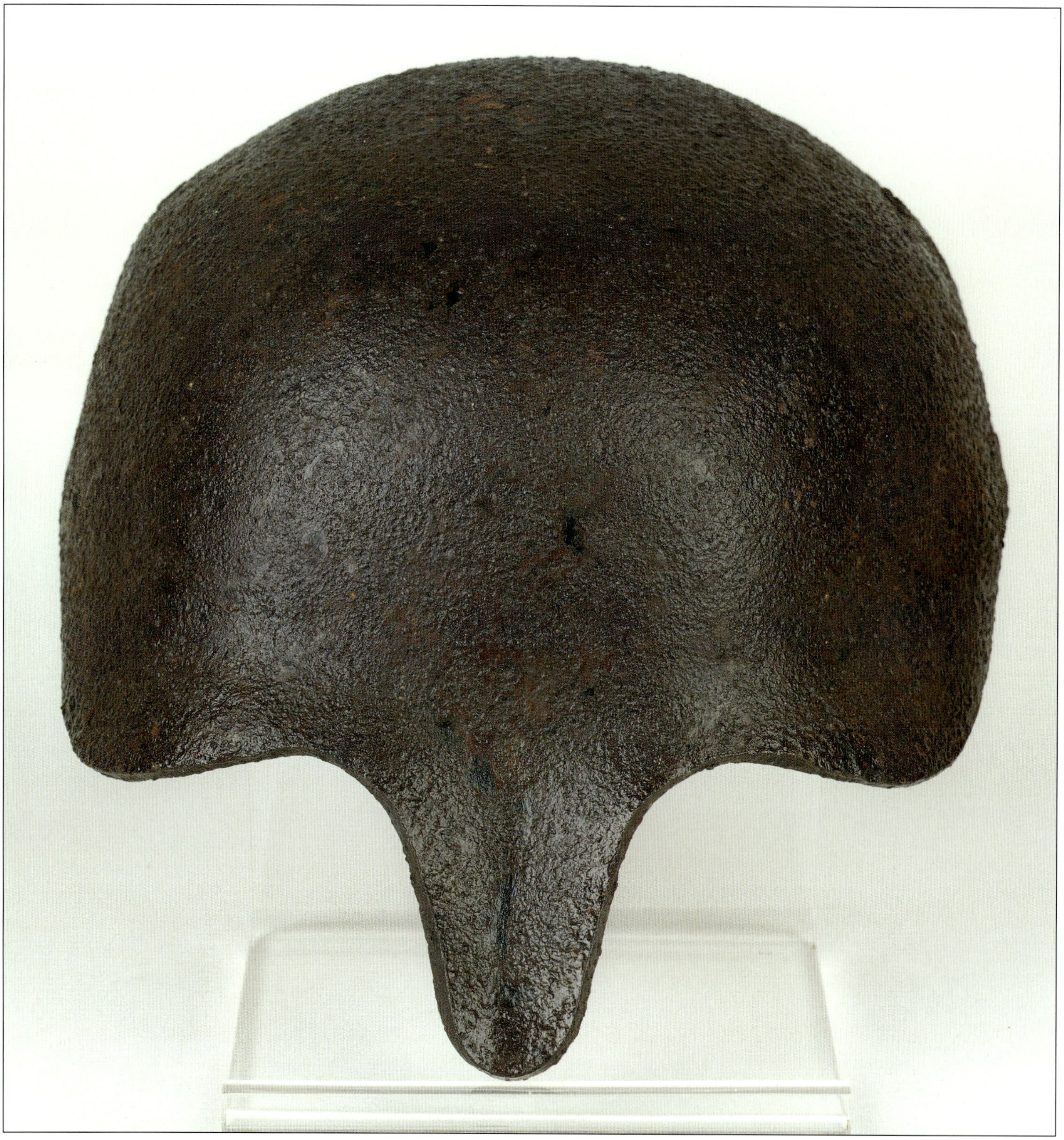

Army Group Gaede Helmet. Relic condition. This example is without the second strap that attaches to the helmet and fastens at the back of the wearer's head, the helmet being kept in place by the skull cap alone. (Collection of Brana Radovic)

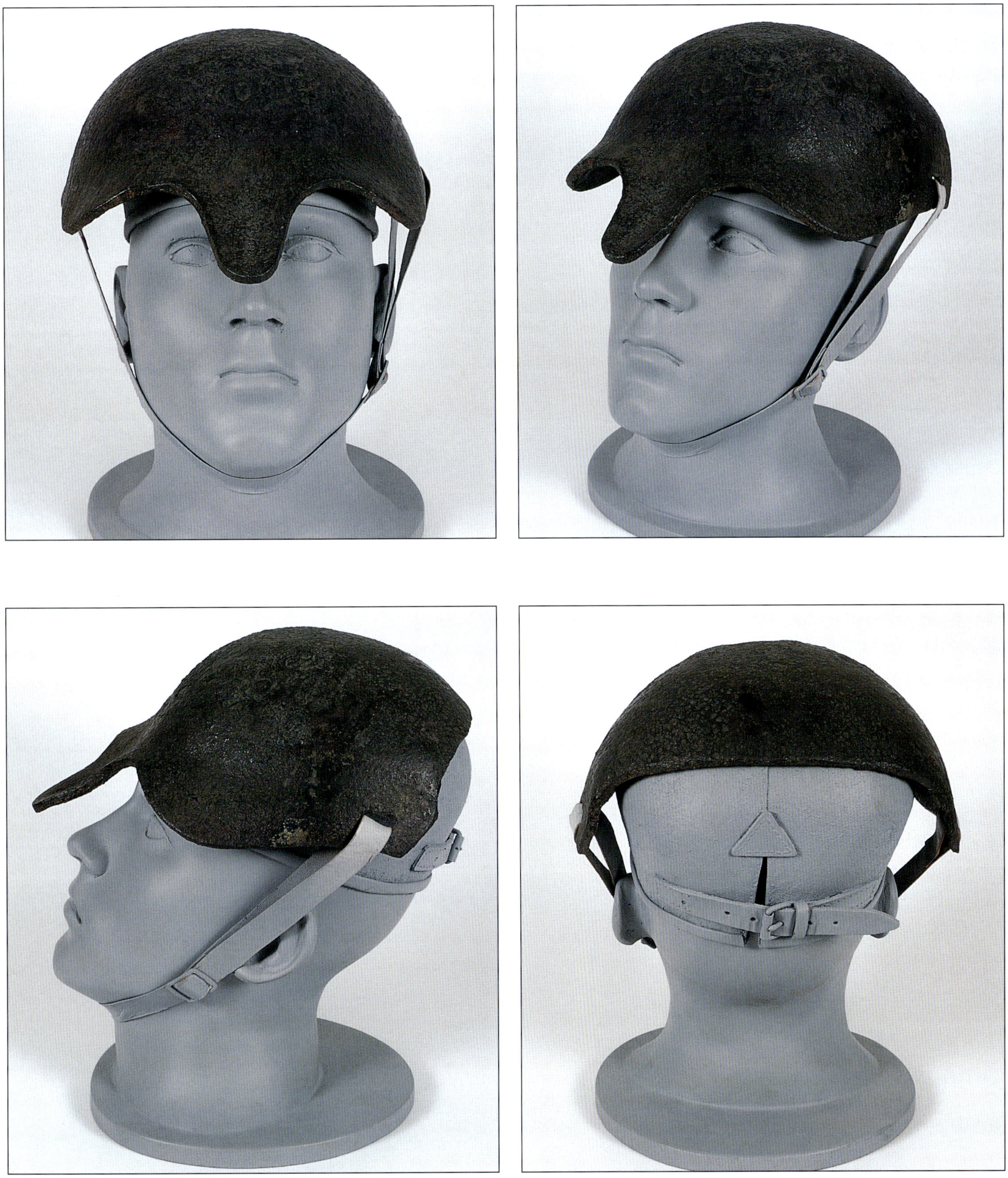

Army Group Gaede Helmet with a reconstructed skull cap and straps. Note that on this example the chinstrap is attached directly to the helmet and not to the skull cap as found in some other versions. Also note that there is no provision for an additional strap across the back of the helmet. (Collection of Michael Baldwin)

This page and following pages. German Facial Defence or Snipers' Face Plate/Helmet. (Collection of Michael Baldwin)

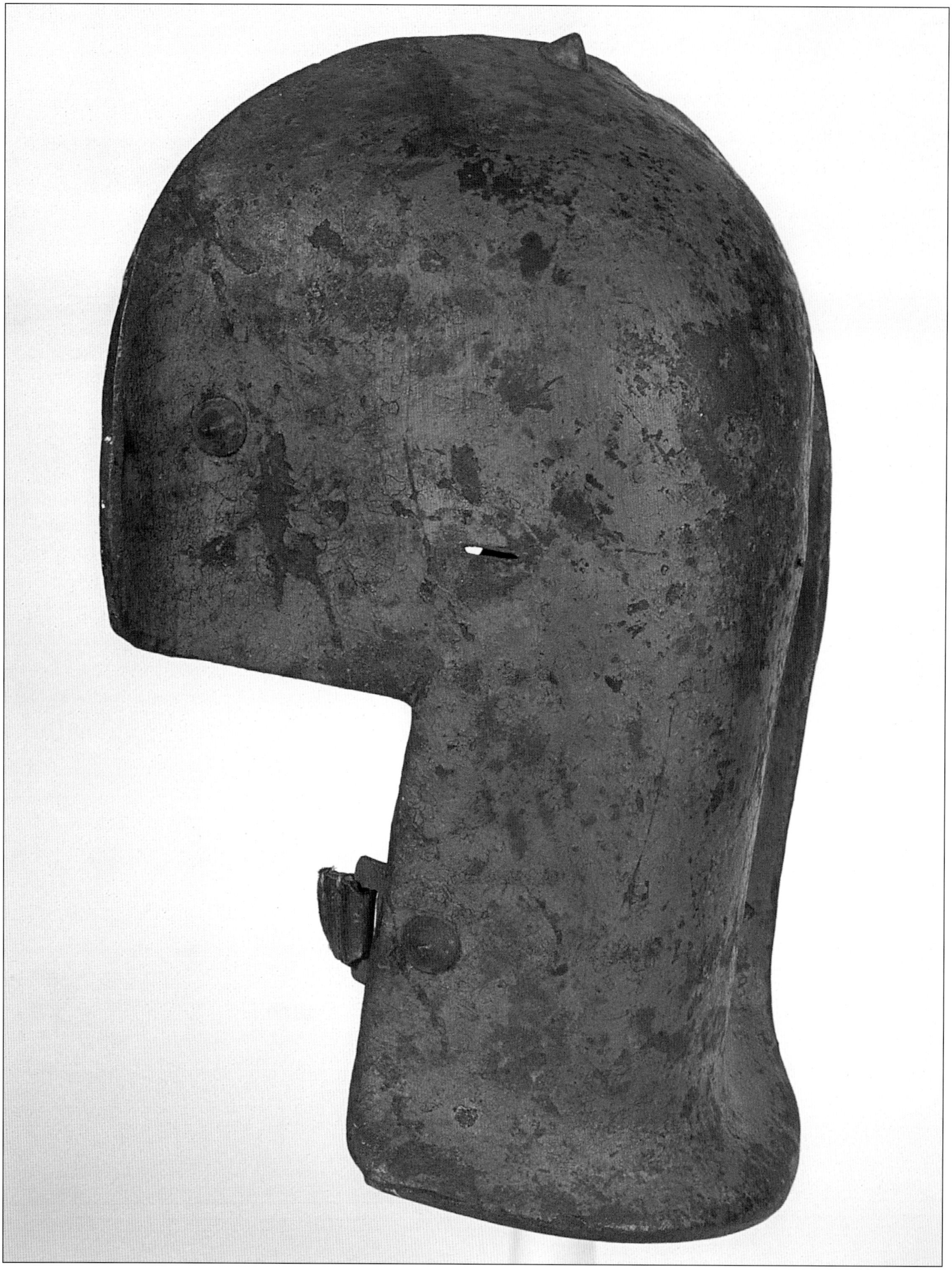

German Model 1916 Helmet (prototype). This example has a "square dip" helmet shell with a light green brushed paint finish. It is fitted with the three-tongued liner pad system (see page 269), pictured in the original patent drawings No. 303460 Class 41c Group 1. (Collection of Brana Radovic)

Four examples of the early version "square dip" German Model 1916 Helmet. (Collections of Marcus Cotton, Imperial War Museum, London, Brana Radovic)

Standard German Model 1916 Helmet. (Collection of Michael Baldwin)

German Model 1918 Helmet. Note absence of chinstrap attachment rivet. (Collection of Brana Radovic)

German Model 1918 cut-out Helmet. Note that it retains some of the original 'flocking" finish and that the ventilation lugs have been blocked. (Collection of Michael Baldwin)

German Model 1916 Helmet variation with additional chinstrap attachments and modified liner and liner attachment. (Collection of Michael Haselgrove)

German Model 1916 Helmet variation with external chinstrap lugs.

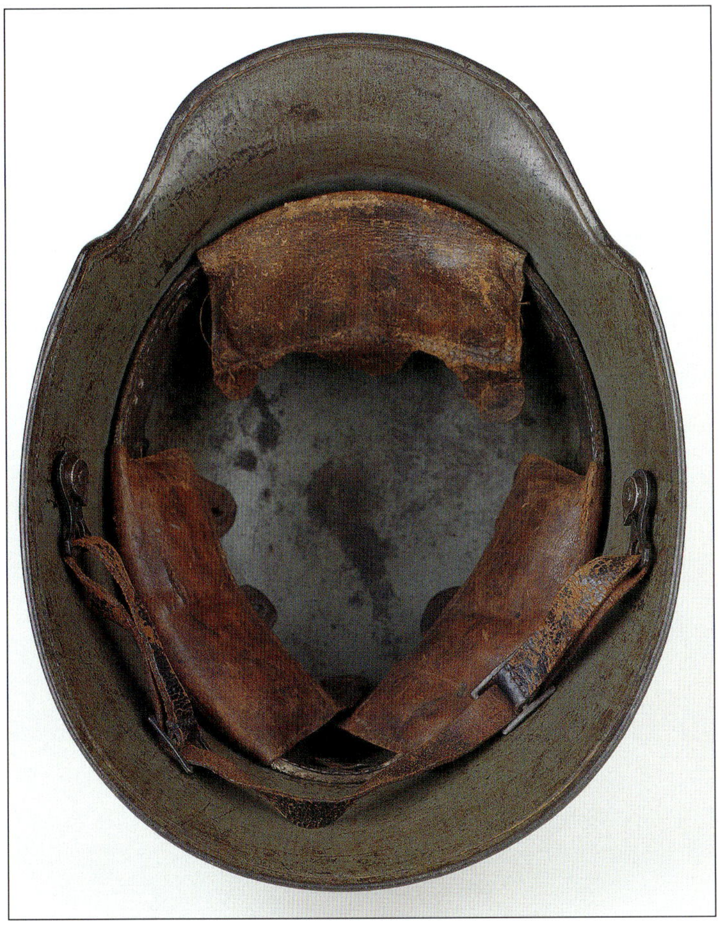

■ Top: Prototype with three tongue liner pad. Note size 66 stamped into the thin leather liner headband often found on "square dip" helmets. (Collection of Brana Radovic)

■ Bottom: Six pad liner variation with individual horse hair stuffed pockets. The authors know of only three examples of this liner type, all fitted to "square dip" helmets. (Collection of Brana Radovic)

German Model 1916 liner. Note liner headband constructed of a double thickness of leather with additional stitching. (Collection of Michael Haselgrove)

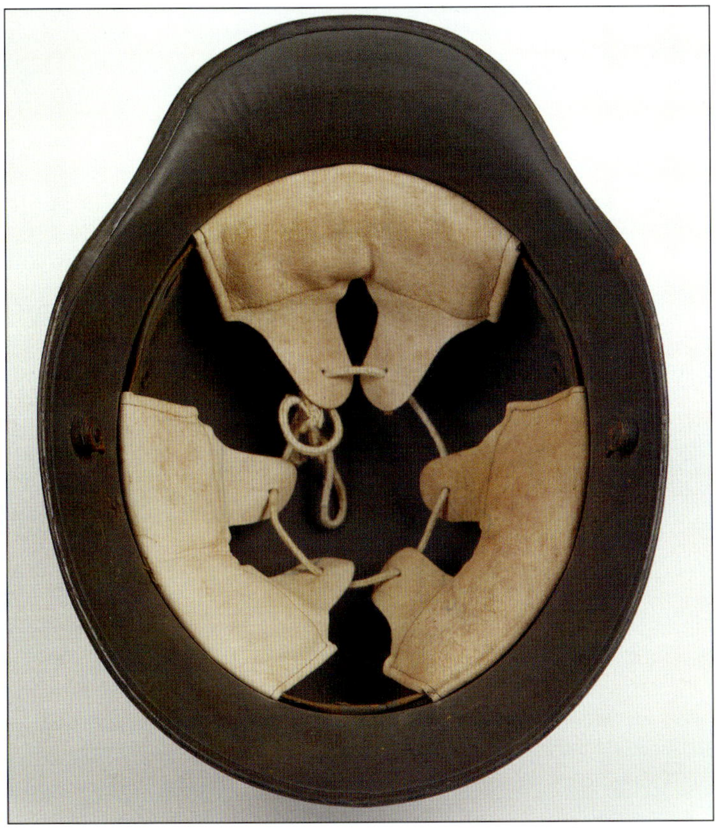

■ Top left: Model 1916 liner with single thickness of liner headband.

■ Top right: Model 1916 liner with single thickness liner headband. Note one row of stitching only.

■ Bottom left: Standard Model 1917 liner with white leather liner pads attached to the metal liner headband. Note size 66 ink stamp to rear of apron.

■ Bottom right: Model 1917 liner with brown leather liner pads. (Collections of Imperial War Museum, London, Michael Haselgrove, Brana Radovic)

- Top left: Standard Model 1918 liner with white leather liner pads.
- Top right: Model 1918 liner with brown leather liner pads.
- Bottom left: Model 1918 cut-out with white leather liner pads.
- Bottom right: Model 1918 cut-out with brown leather liner pads.

Ventilation lug sizes. Top left: Size 60. Top right: Size 62. Bottom left: Size 64. Bottom right: Size 66-68. Note the rounded end to the size 64 ventilation lug.

Manufacturers' marks and sizes. Top row: Model 1916. Right: Note hammered lug preventing loss of chinstrap attachment. Middle row: Model 1917. Bottom row: Model 1918. Note twin rivets securing chinstrap.

Examples of Model 1916/17 chinstraps. Note the ersatz replacement on right. (Collection of Michael Baldwin)

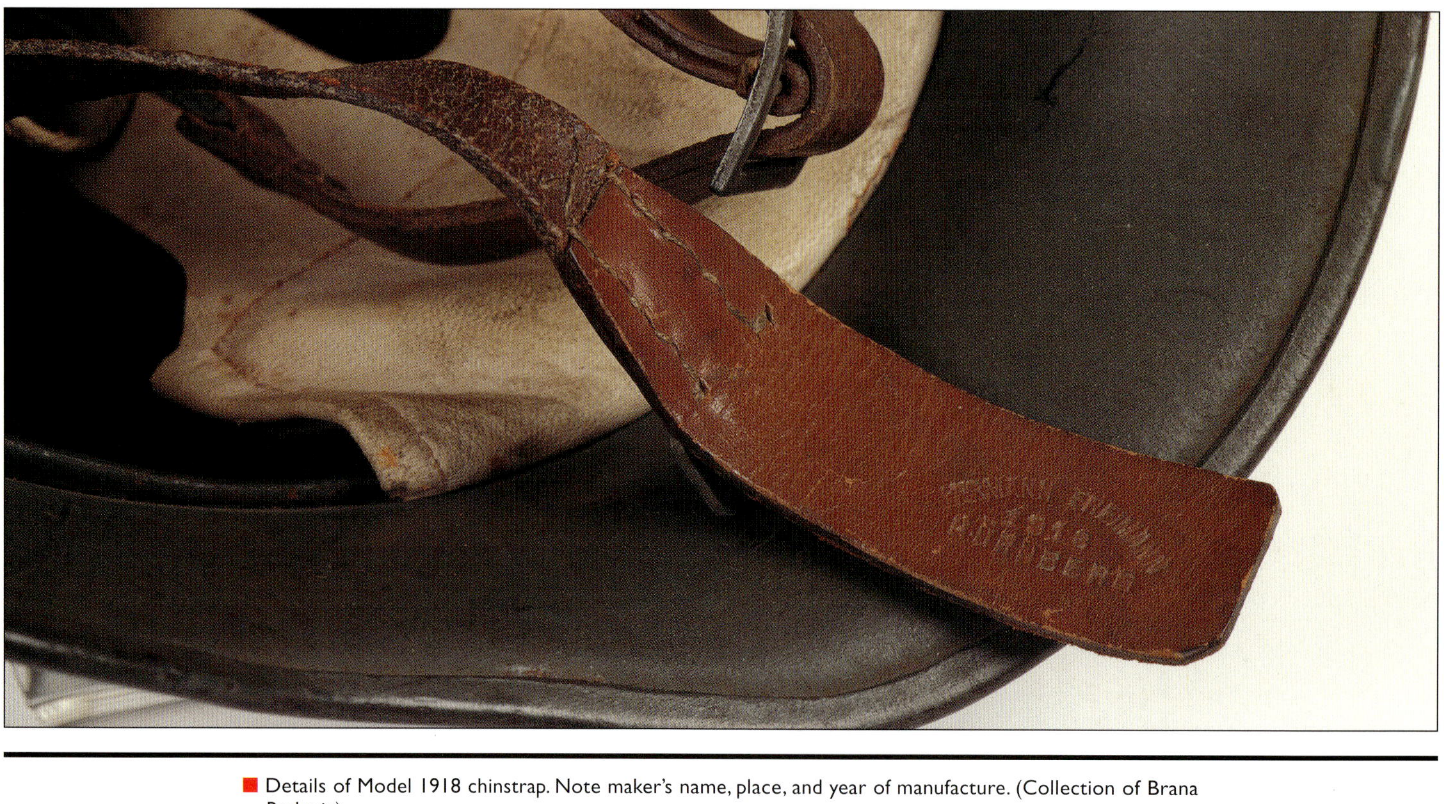

Details of Model 1918 chinstrap. Note maker's name, place, and year of manufacture. (Collection of Brana Radovic)

Top: Interior view of Model 1916 liner. Note the manufacturer's details on the headband.

Bottom: Interior view of Model 1917 liner. (Collection of Michael Baldwin)

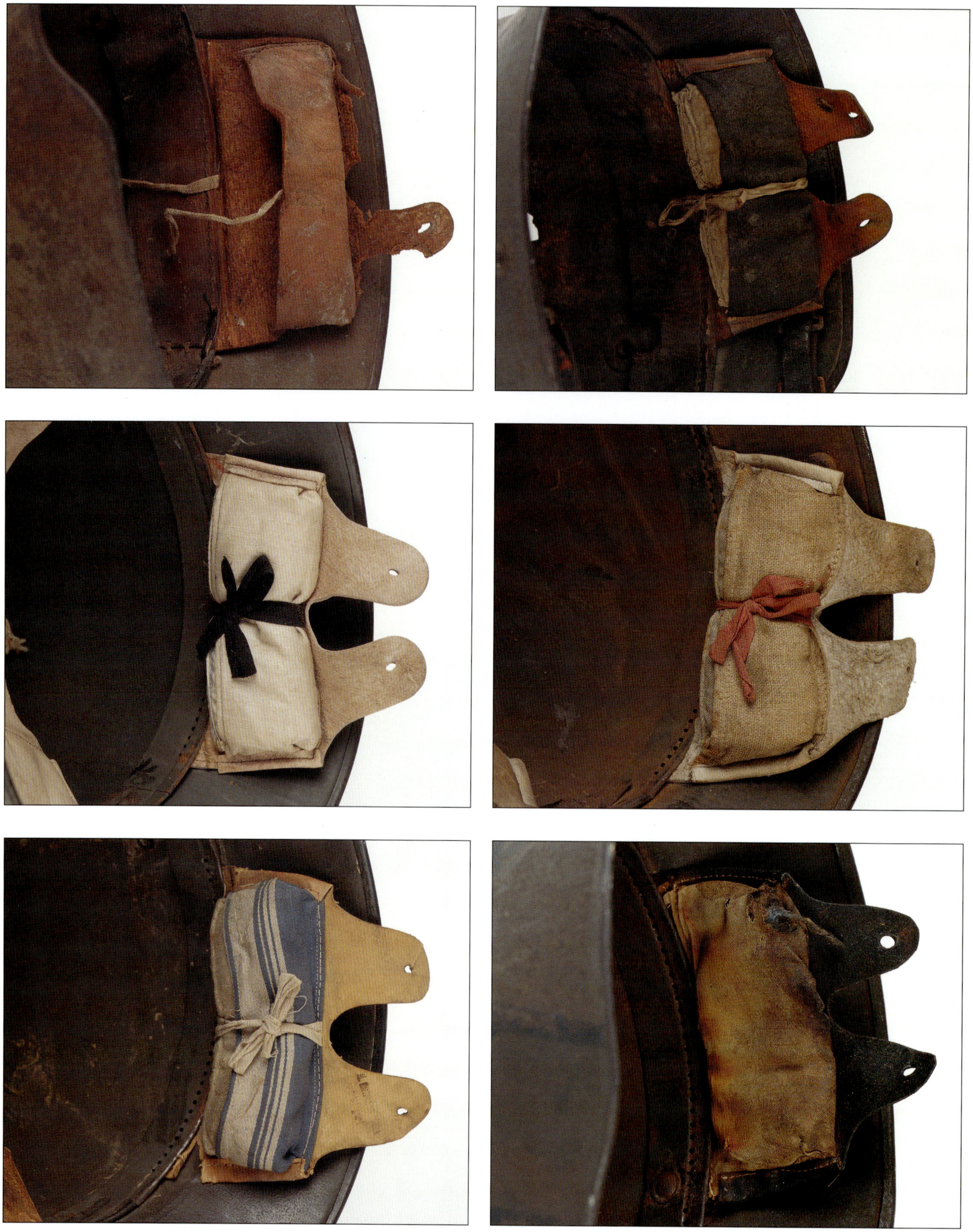

Examples of liner pads and pad pockets. Note top left: The early liner pad pocket is formed from leather instead of cloth. Note also bottom right: There is no liner pad pocket, the padding being stitched into place.

Six examples of liner pad cushions.

German Model 1916/17 Helmet. This camouflaged helmet complies with the Ludendorf Directive and has wide evenly applied black separation lines that some say indicate a factory-applied finish. However it may be that these well produced paint schemes are the result of the work of skilled, creative members of the German Army. (Collection of Michael Baldwin)

German Model 1916/17 Helmet. The significance of the red painted cross on the front is unknown, but is possibly medical. (Collection of John Burnett)

German Model 1916/17 Helmet. (Collection of Brian Maederer, Military Collectibles)

German Model 1916/17 Helmet.

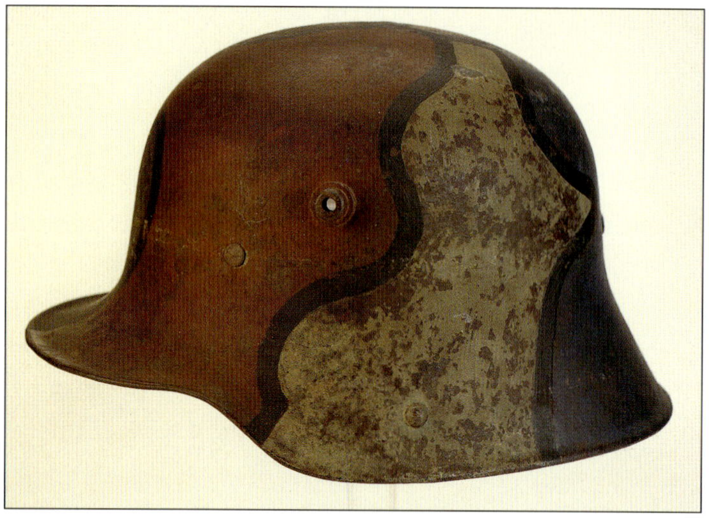

■ These pages and the following two pages. German Model 1916/17 Helmets. (Collections of John Burnett, Robert Tredwen, Brian Maederer, Brana Radovic, Marcus Cotton, Michael Baldwin, David Acon, Michael Haselgrove, Ellwood von Seibold, Imperial War Museum, London)

German Model 1918 Helmet. (Collection of Michael Baldwin)

German Model 1918 Helmet. (Collection of Michael Haselgrove)

German Model 1918 Helmet. Note bird shape incorporated into paint scheme (Collection of Brana Radovic)

German Model 1918 cut-out Helmet. (Collection of Michael Baldwin)

German Model 1918 cut-out Helmet. (Collection of Malcolm Fisher, Regimentals)

German Model 1918 cut-out Helmet. (Collection of John Burnett)

Top: German Model 1916/17 Helmet. Note. Machine gunners badge applied to the right side. (Collection of Brana Radovic)

Bottom: German Model 1916/17 Helmet. Note. Machine gunners badge applied to the left side. This helmet also has a Prussian eagle, in the same paint, on the front. (Collection of John Burnett)

■ Top: German Model 1918 Helmet. Note that his helmet has what appears to be a small Hesse shield on the left side. It is usually assumed that painted shields of this type were only applied post-Great War. However photographic evidence for their use in the Great War can be found in the period photographic section. (Collection of Brana Radovic)

■ Bottom: German Model 1916/17 Helmet with Prussian 1st Guarde Regiment shield applied to the left-hand side. This helmet also has a dark painted company number "4" on the rear centre, partially covered by the remains of an adhesive plaster post-home label.

■ Camouflaged pattern drawstring cover, possibly made from German aircraft fabric. (Collection of Brana Radovic)

Regulation pattern 7-clip cloth cover. (Collection of Brana Radovic)

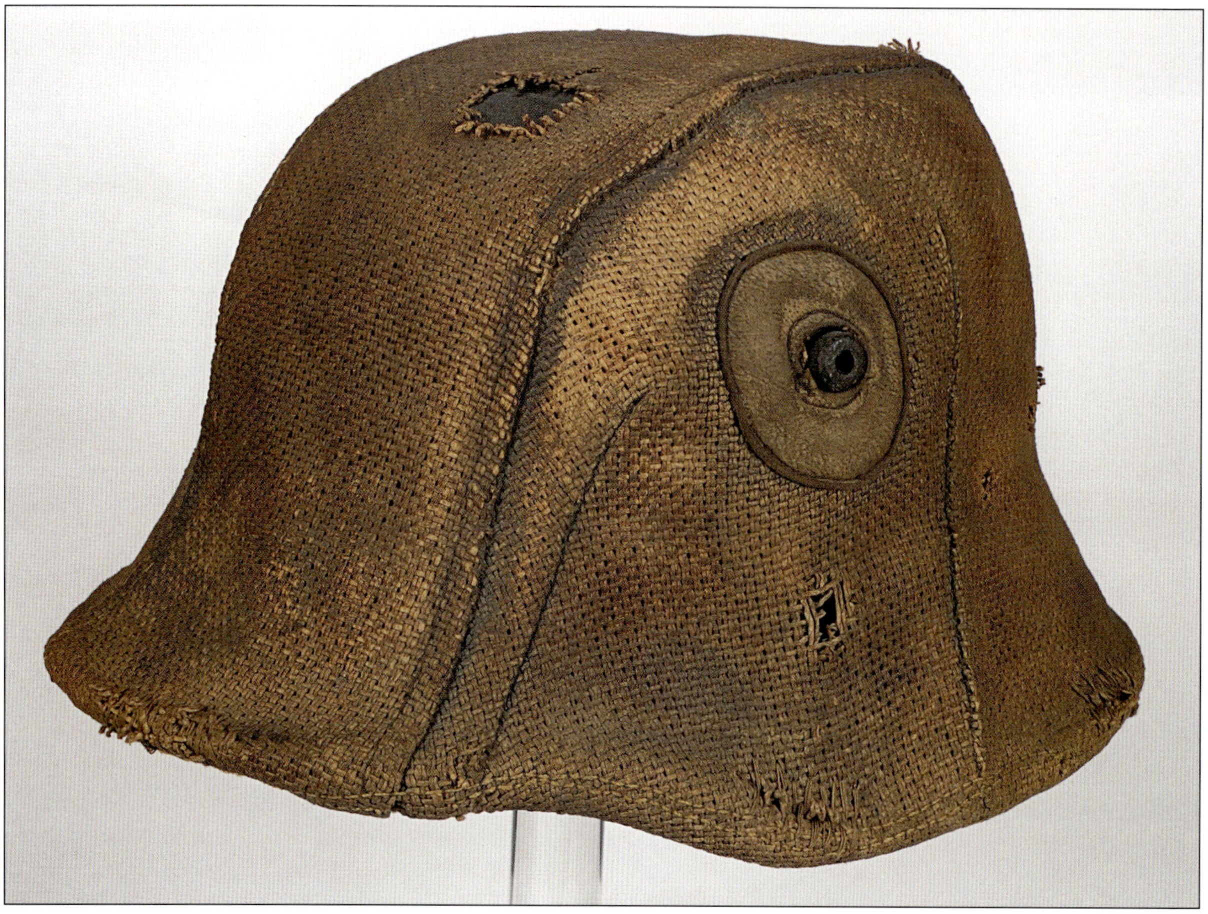

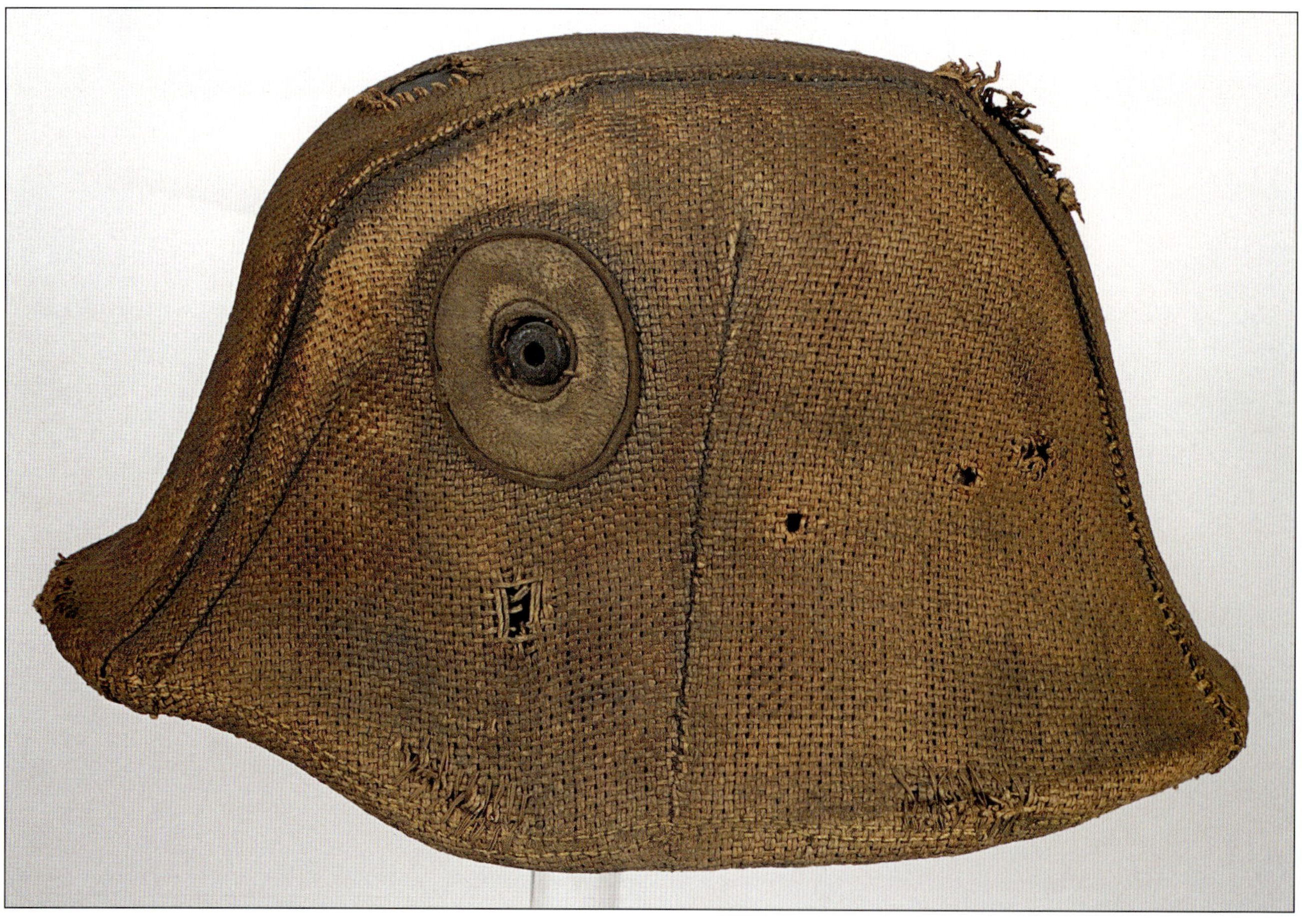

Non-regulation woven hessian drawstring cover. (Collection of Brana Radovic)

German Model 1918 Helmet with permanently attached cloth cover, secured by heavy wire. (Collection of Brana Radovic)

German Model 1916/17 Helmet with a Pickelhaube cover. The cover has been adapted by the addition of twisted cloth extensions that allow the clips to reach the helmet rim. Note that the Pickelhaube chinstrap slits are utilised for the ventilation lugs. A painted camouflage finish has then been applied over the helmet cover and helmet. (Collection of Imperial War Museum, London)

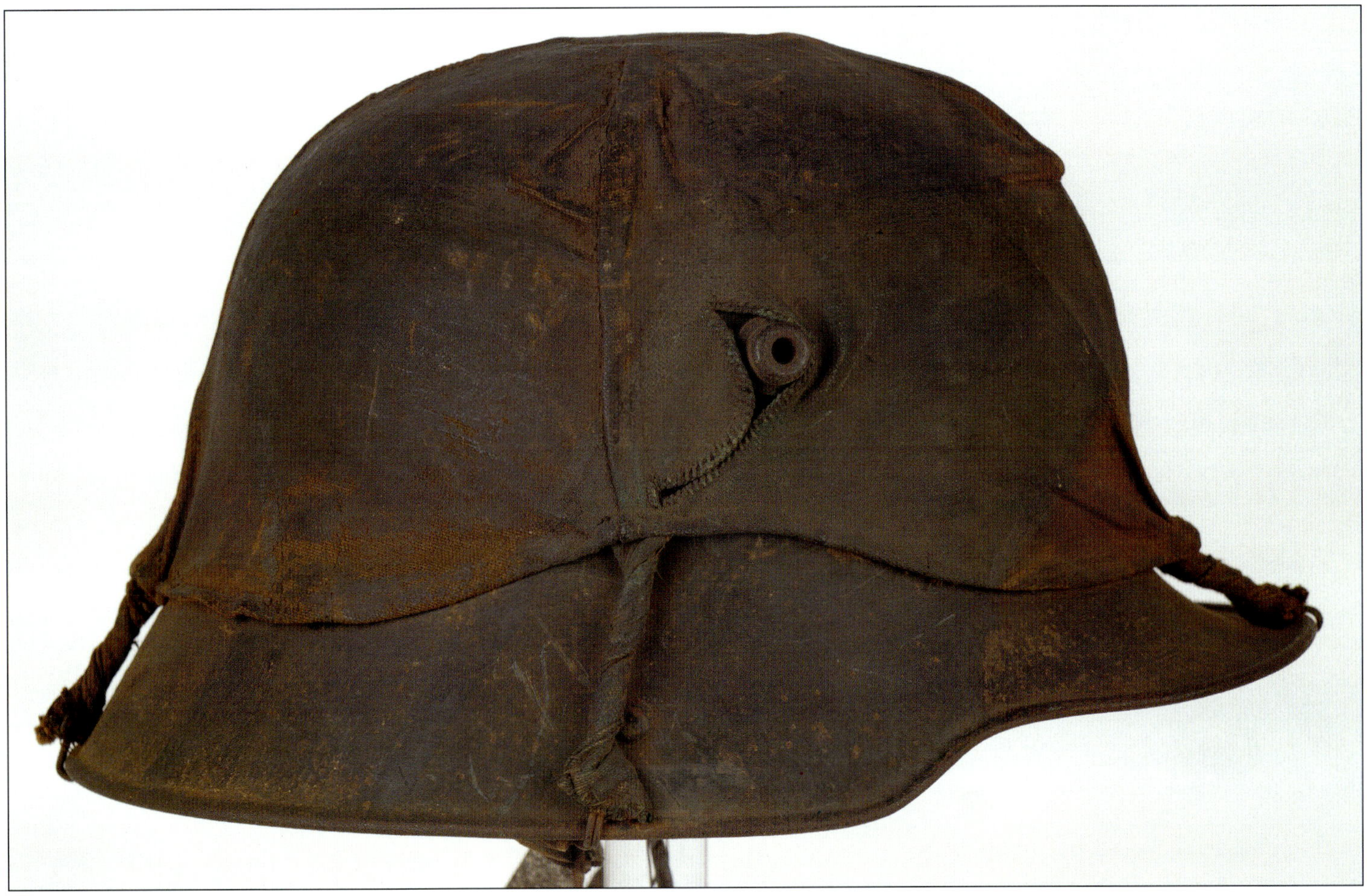

German brow plate. This example has a painted Iron Cross on the front. (Collection of Imperial War Museum, London)

Another example of the brow plate. Note the two rivets securing the strap and the superfluous holes punched through the leather to take the buckle. (Collection of Robert Tredwen)

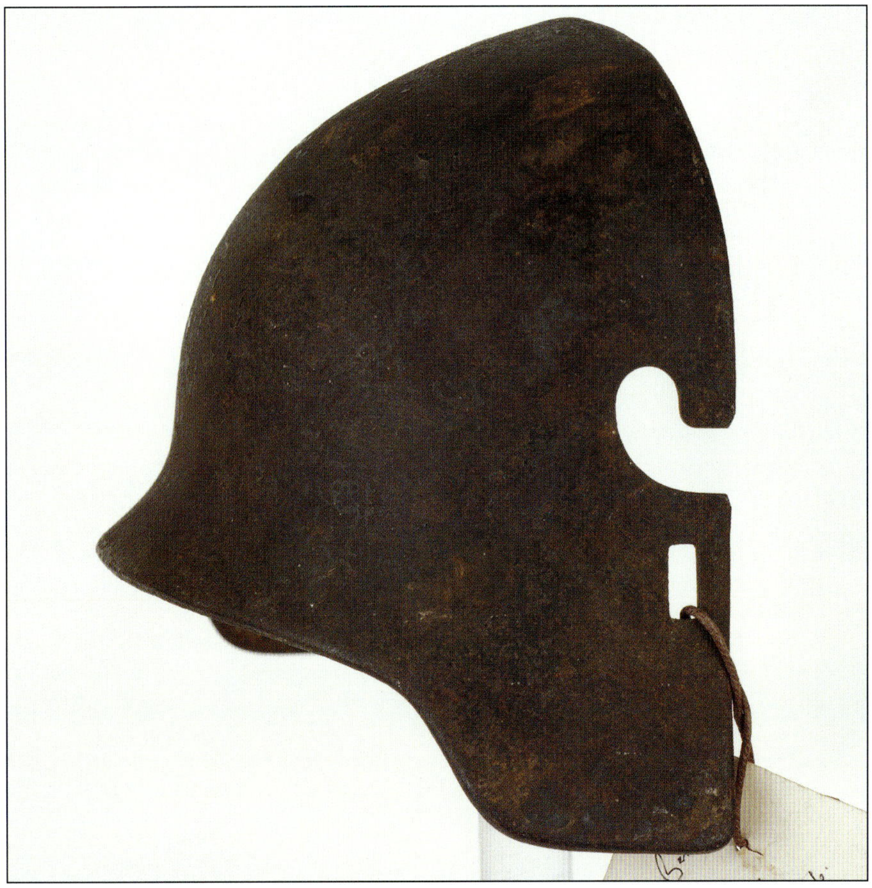

Another example of the brow plate. The label reads "Forwarded by L. Col A.F. Burt MC. Taken by my brother CAP. J Burt M.G.C at Mametz Somme 1916. He was killed near Amiens in May 1918". (Collection of Imperial War Museum, London)

Examples of battle-damaged helmets. (Collections of Michael Haselgrove, Imperial War Museum, London, Brana Radovic)

Souvenired Model 1916/17 Helmet. With the inscription "Bapaume 1919". (Collection of Marcus Cotton)

Souvenired Model 1916/17 Helmet. With the inscription " 12th Div. FRANCE, 1915–19. With depictions of various regimental badges of units serving with the 12th Division. This helmet was presented to General Scott GOC 12th Div. (Collection of Marcus Cotton)

THE NORFOLK
ROYAL ENGINEERS
ROYAL WEST KENT

Souvenired Model 1916/17 Helmet. With battlefield scenes and views of France. (Collection of Ellwood von Seibold)

■ Top: Camouflaged model 1916 "square dip helmet" with American postage stamps attached. (Collection of Brana Radovic)

■ Bottom: Model 1916 helmet with souvenir label attached to the liner headband. The label reads: Hun "Coal scuttle" Found along a Sunken road near dead Bodies. YPRES SECTOR Dec 1917. (Collection of Michael Haselgrove)

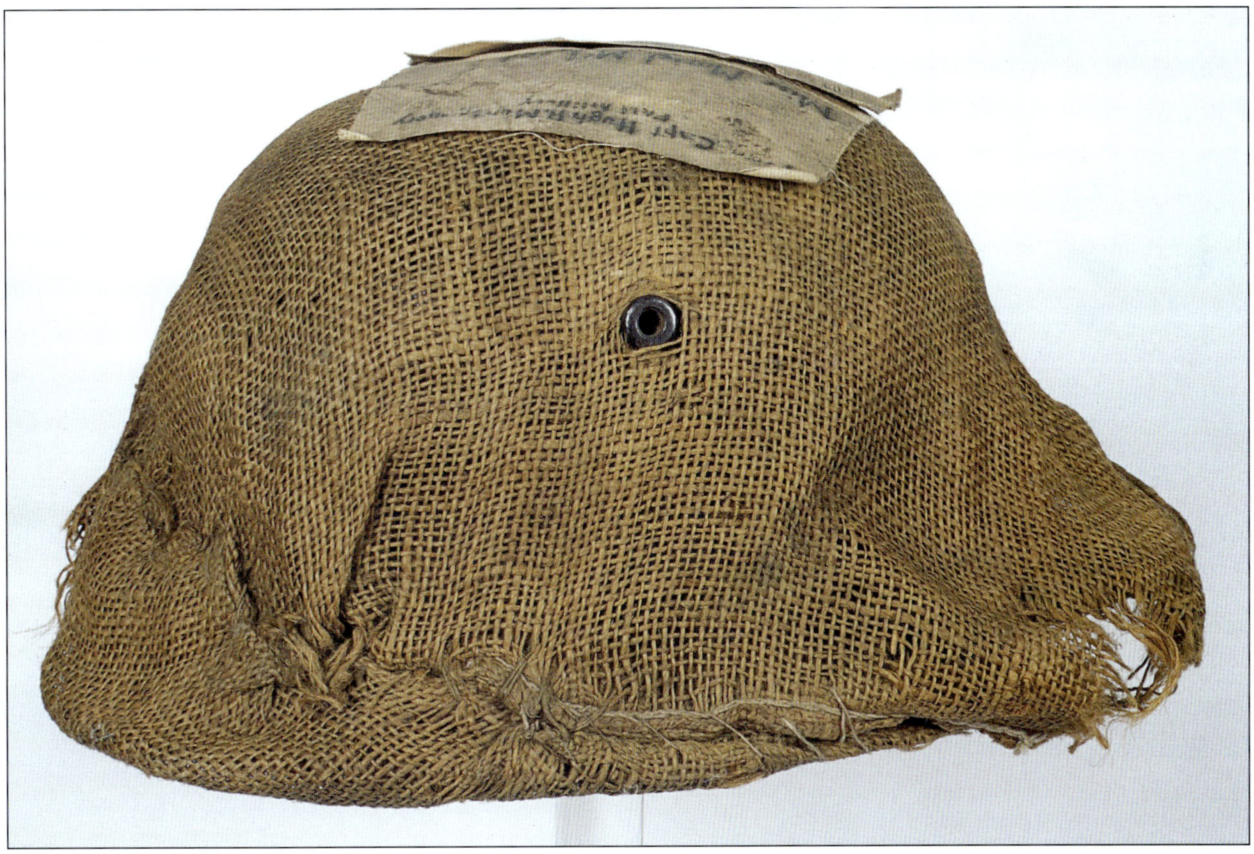

■ Top: Camouflaged model 1916/17 Helmet with American post-home label. The label reads "WAY BEN CLAYK AMB Co 125 AEF. HP073H MISS MAYIEETTA. CLAYK PEWAUKEE WISCONSIN USA OK AMB. Co 125" (Collection of Brian Maederer, Military Collectibles)

■ Bottom: Battle damaged model 1916/17 Helmet and cover. The woven hessian cover bears a post-home label. (Collection of Robert Tredwen)

Trench art coal-scuttle. (Collection of Brana Radovic)

■ Top: The process of creating a helmet. From a flat circular sheet of steel, through the seven stamping processes and thereafter to the completed helmet.

■ Bottom: Surviving examples of the stampings in the Dresden Museum, Germany.

Top: A photograph of the newly introduced Model 1916 helmet in use. Note the square dip and that the soldiers have painted their regimental number to the front. Bundesarchiv

Bottom: Photograph of troops in training wearing Model 1916 square dip helmets. Budesarchiv

Note the Model 1916 square dip helmets. Greg Nevill

The Model 1916 helmet

The Model 1916 helmet. Stephen Chambers

The Model 1916 helmet

The Model 1916 in use on the Western Front. Bundesarchiv

■ Top: Model 1916 helmet

■ Bottom: Model 1918 helmet

Top: Assault troops. Stephen Chambers

Bottom: No caption available. The Model 1918 cut out in use. This photograph was taken shortly after the war, probably in Germany in 1919. IWM Q110476

A unit specialising in the camouflaging of helmets (and possibly other objects).

Top: Troops moving artillery pieces. They are wearing camouflaged helmets. Bundesarchiv

Bottom: A light gun being taken forward in sections on the shoulders of the personnel. Note the camouflage helmets and added shield insignia. IWM Q56514

Top: Battle of the Canal du Nord. 4th Canadian Division. Wounded and prisoners passing through a Field Dressing Station. Near Moeuvres, 27 September 1918. Note the camouflage helmets. IWM Q9331

Bottom: Happy though captives. German prisoners bagged by Australians in a recent stunt on the Somme. 22.8.18. near Bray. 34TH & 35TH BATTALIONS A.I.F. Note camouflage helmets. IWM E(AUS)2949

Top: Battle of Amiens. Prisoners arriving at a 'cage' near Amiens. 9 August 1918. Note camouflage helmets. IWM Q9193

Bottom: Prisoners and wounded coming back. A Canadian photograph showing camouflage German helmets. IWM CO3310

■ Top: Prisoners and wounded passing 2nd Canadian Siege Battery. Note camouflage German helmets. IWM CO3311

■ Bottom: Battle of Albert. Prisoners bringing in wounded and captured machine-guns. Near Courcelles, 21 August 1918. Note camouflage German helmets. IWM Q6968

Top: Camouflaged helmet in use.

Bottom: Training photograph of flamethrower troops wearing cloth helmet covers. Bundesarchiv

■ Top: MG08/15 in use during training. Note different styles of cloth helmet covers being used (with and without provision for the ventilation lugs). IWM Q23678

■ Bottom left: Wounded arrive at a Canadian Field Dressing Station. IWM CO2951

■ Bottom right: Sentry wearing what appears to be a clip-on cloth helmet cover.

Note cloth helmet covers

Unknown type of armoured frontal protection. Judging from the style of grenade, the photograph was probably taken in 1915 or 1916. Stephen Chambers

German armoured face shield.

Top and bottom: A German sniper's mask made of 1/2" Krupp steel. IWM CO1865/1866

Top and bottom: German artillery observation post. Note the helmet with armoured brow plate

Brow plates being worn. Stephen Chambers/Bundesarchiv

Top: Salvaging steel helmets captured from the enemy. 18 January 1918. IWM Q8444

Bottom: War trophies captured by the Australians during the Battle of Hamel. 12.7.18. Ailly-sur-Somme. AUST. WAR. RECORDS. Note the brow plate at centre. IWM E(AUS)2743

Reserve Infantry Regiment 120, Cambrai. Note some soldiers are wearing body armour. Bundesarchiv.

Ambushing an English patrol, La Vacquerie August 1917. Bundesarchiv.